Autonomous Nature

In this ambitious history of ideas, Carolyn Merchant calls attention to the ancient idea of nature as unpredictable, rebellious, and impossible to understand and control completely. She urges us to recover that older idea for the foundation of a new ecological ethic. Wide-ranging, original, and provocative.

Donald Worster, author of "Nature's Economy:
A History of Ecological Ideas"

Merchant has written a key history of ideas for evaluating two of the big questions today: how did we get into this mess, and how can we get out of it. Western thinkers, who gave us the scientific method, also fell short of the truer, fuller view of reality, dynamical and chaotic. It is against this richer backdrop that we can grasp today's emerging complexity paradigm, and find hope and insight for restoring our planet's beautifully 'rambunctious gardens.'

Jennifer Wells, California Institute of Integral Studies,
author of Complexity and Sustainability

Autonomous Nature investigates the history of nature as an active, often unruly force in tension with nature as a rational, logical order from ancient times to the Scientific Revolution of the seventeenth century. Along with subsequent advances in mechanics, hydrodynamics, thermodynamics, and electromagnetism, nature came to be perceived as an orderly, rational, physical world that could be engineered, controlled, and managed. *Autonomous Nature* focuses on the history of unpredictability, why it was a problem for the ancient world through the Scientific Revolution, and why it is a problem for today. The work is set in the context of vignettes about unpredictable events such as the eruption of Mt. Vesuvius, the Bubonic Plague, the Lisbon Earthquake, and efforts to understand and predict the weather and natural disasters. This book is an ideal text for courses on the environment, environmental history, history of science, or the philosophy of science.

Carolyn Merchant is Professor of Environmental History, Philosophy, and Ethics at the University of California, Berkeley. She is the author of The Death of Nature; Ecological Revolutions; and Reinventing Eden among other books. She is a past president of the American Society for Environmental History and a recipient of the Society's Distinguished Scholar Award.

Related Titles from Routledge

Reinventing Eden: The Fate of Nature in Western Culture, Second Edition
Carolyn Merchant

Off the Grid: Re-Assembling Domestic Life
Phillip Vannini and Jonathan Taggart

Sacred Ecology, Third Edition
Fikret Berkes

Autonomous Nature
Problems of Prediction and Control from
Ancient Times to the Scientific Revolution

Carolyn Merchant

NEW YORK AND LONDON

First published 2016
by Routledge
711 Third Avenue, New York, NY 10017

and by Routledge
2 Park Square, Milton Park, Abingdon, Oxon, OX14 4RN

Routledge is an imprint of the Taylor & Francis Group, an informa business

© 2016 Taylor & Francis

The right of Carolyn Merchant to be identified as author of this work
has been asserted by her in accordance with sections 77 and 78 of the
Copyright, Designs and Patents Act 1988.

All rights reserved. No part of this book may be reprinted or reproduced or
utilised in any form or by any electronic, mechanical, or other means, now
known or hereafter invented, including photocopying and recording, or in
any information storage or retrieval system, without permission in writing
from the publishers.

Trademark notice: Product or corporate names may be trademarks or
registered trademarks, and are used only for identification and explanation
without intent to infringe.

Library of Congress Cataloging-in-Publication Data
Merchant, Carolyn.
   Autonomous nature : problems of prediction and control from ancient
times to the scientific revolution / by Carolyn Merchant.
      pages cm
   Includes bibliographical references and index.
   1. Philosophy of nature.   2. Chaotic behavior in systems.
3. Complexity (Philosophy)   4. Nature—Forecasting.   5. Nature—
Effect of human beings on.   I. Title.
      BD581.M395 2015
      113—dc23
      2015012689

ISBN: 978-1-138-93099-5 (hbk)
ISBN: 978-1-138-93100-8 (pbk)
ISBN: 978-1-315-68000-2 (ebk)

Typeset in Minion
by Apex CoVantage, LLC

For David and John

Contents

List of Figures and Tables	ix
Preface	xi
Introduction: Can Nature Be Controlled?	1

PART I
Autonomous Nature — 19

1	Greco-Roman Concepts of Nature	21
2	Christianity and Nature	42
3	Nature Personified: Renaissance Ideas of Nature	63

PART II
Controlling Nature — 79

4	Vexing Nature: Francis Bacon and the Origins of Experimentation	81
5	Natural Law: Spinoza on *Natura naturans* and *Natura naturata*	101
6	Laws of Nature: Leibniz and Newton	125
	Epilogue: Rambunctious Nature in the Twenty-First Century	149

Bibliography	169
Index	183

List of Figures and Tables

Figures

1.1	Vesuvius in Eruption, painted ca., 1776–1780	22
2.1	1348 Earthquake	43
3.1	Black Death, 1348	64
4.1	Demonic Vexation, 1597	82
4.2	Francis Bacon (1561–1626)	83
4.3	Robert Boyle (1627–1691)	91
4.4	Robert Boyle's Air Pump, 1744	92
4.5	An Experiment on a Bird in the Air Pump, 1768	93
5.1	Baruch Spinoza (1632–1677)	102
6.1	Spinoza's House in The Hague Where He Lived in 1676	126
6.2	Gottfried Wilhelm Leibniz (1646–1716)	127
6.3	Isaac Newton (1642–1727)	128
6.4	The Great Plague of London, 1665	133
6.5	The Lisbon Earthquake, 1755	142

Table

1.0	Meanings of Nature	12

Preface

Autonomous Nature stems from a desire to understand the roots of the idea of nature as an active, sometimes disruptive and unruly entity in the past and present. Manifested most dramatically in the effects of climate change, natural disasters, habitat transformations, species dislocations, and the loss of human homes and lives, such disruptions result from ecological tipping points and cascading effects that are often unpredictable and uncontrollable.

This book is intended for those with a background in the humanities, social sciences, and the sciences and for an educated audience interested in the past, present, and future of humanity and life on earth. It proposes that the new sciences of chaos and complexity theory not only constitute a new paradigm for twenty-first century science, but are the basis for a new ethic of partnership with a nature that must be considered as autonomous and often unpredictable. To understand how humanity has arrived at this new paradigm and environmental ethic, this book investigates the early history of nature in Western culture.

Autonomous Nature explores the idea of nature in the Western world from ancient times to the (so-named) Scientific Revolution of the seventeenth century as an active, chaotic, unruly, and unpredictable actor/actress and the ways in which natural scientists and philosophers sought to control and manage it for human benefit. It looks at the tensions between order and chaos, stability and change, predictability and unpredictability. It delves into well-known texts of the past, looking, in particular, for what they said about disruption, disorder, and disaster and how to deal with the uncertainties of nature, both philosophically and in the everyday world. In so doing, it reframes and reinterprets many well-known works, focusing especially on ancient and early modern European culture.

The book draws on familiar authors and events, but its goal is to reread them in ways that shed new light on twenty-first century problems, in particular the often unanticipated effects of natural disasters and climate change. Using the new sciences of the late twentieth century, as guidelines and lenses, it looks at the history of the unruliness and unpredictability of nature in the early texts of Western culture. It does not intend to cover all individuals or events equally or to deal with non-Western cultures, but rather to select those texts that shed light on the prehistory of chaos and complexity theory. And while many histories of science

xii *Preface*

presuppose progress, order, and the emergence of new ideas, this book looks at ideas of disorder and the ways that natural philosophers sought to subdue it.

Since the early 1990s I have been writing about nature as an autonomous actor and humanity in partnership with nonhuman nature. At the Earth Summit in Rio de Janeiro in 1992, I formulated what I call a "partnership ethic" which holds that "the greatest good for the human and the nonhuman community is to be found in their mutual living interdependence." (See "Conclusion: Partnership Ethics," in my 1996 book, *Earthcare*.)

My book also goes back to a question that arose during the 1970s as I was writing *The Death of Nature* (1980). What were the origins and influences of the terms *natura naturans*, or nature creating, and *natura naturata*, the created world? Those concepts were discussed by R. G. Collingwood in his 1939 book, *The Idea of Nature*, and by Eustace Tillyard in *The Elizabethan World Picture* (1959). In *The Death of Nature*, I wrote about the idea that nature, depicted as a female during the Renaissance, seemed not only to be the instrument of God, but also to exhibit a will of "her" own. Since then, as I have investigated the origins of chaos and complexity theories that emerged in the 1970s and 1980s, I have returned to the history of nature depicted as willful, unlawful, and hence unpredictable. In addition to *The Death of Nature*, I referred to *natura naturans* and *naturata* in several places including *Ecological Revolutions* (1989) and *Reinventing Eden* (2002). I also raised the question of the origin of these two terms in conversations with historian of medieval science Amos Funkenstein in the 1990s, who promptly answered, "it was *natura creans* and *natura creata*" and philosopher J. Baird Callicott who said, "Spinoza used them, but in a different way."

In 2012–13, I received a fellowship from the American Council of Learned Societies to study the topic, "Ideas of Nature: Emerging Concepts of Law and Nature in the Scientific Revolution," that focused on the history and meaning of those two concepts. I am deeply grateful for their support. I was also invited to work on the project at Princeton's Institute for Advanced Study (IAS) during the fall term of 2012 with support from the Mellon Foundation. At IAS I am indebted to Jonathan Israel, Irving and Marilyn Lavin, Hyun Ok Park, Anne-Lise Rey, Frans van Liere, and especially to Heinrich von Staden for conversations, references, and insights and to the staff of the Institute for library and housing support. At Princeton University I wish to thank Daniel Garber and the university's librarians. In addition, I very much appreciate the response of Brother Alexis Bugnolo in Rome to my inquiries regarding Saint Bonaventure's *Commentaries on the Sentences of Peter Lombard* and for his translations of the Latin passages discussed in the notes to Chapter 2 on "Christianity and Nature."

I am especially indebted to Francesca Rochberg with whom I collaborated on a UC Berkeley Townsend Center for the Humanities Collaborative Research Seminar entitled "Nature/No Nature: Rethinking the Past, Present, and Future of Nature in the Contemporary Humanities." During the spring semester of 2012, together with faculty colleagues and graduate students in the humanities and sciences, we explored the origins and meanings of nature, focusing primarily on the history of Western culture. I am indebted to Francesca's inspiration and support as well as to the seminar members for their insightful and scholarly contributions and for the support of the Townsend Center. The University of California at Berkeley awarded me sabbatical leave during 2012 and the Committee on Research provided me with a Faculty Research Grant (FRG) and a Humanities Graduate Student Research (GSR) Grant for 2012.

I also wish to thank Sheila ffolliott for her references and insights into Renaissance art and David Kubrin for numerous conversations over the past years and references on seventeenth-century science and society. Robert Westman offered many insights into reappraising the issue of the "scientific revolution." Jennifer Wells shared her ideas about complexity theory and her book on *Complexity and Sustainability* (2013) as it has evolved over the past few years. Ken Worthy engaged in conversations with me about *natura naturans* and shared ideas for his book on *Invisible Nature* (2013). I am particularly indebted to all those who read chapters of the manuscript for their astute comments and helpful suggestions for improving the argument. I especially thank Celeste Newbrough for the index to this and previous books and to Ted Grudin for his suggestions and careful reading of the manuscript.

Thank you also to the Routledge reviewers: Glenn Fieldman, San Francisco State University Yan Gao, University of Memphis Jennifer Wells, California Institute of Integral Studies Marco Armiero, Royal Institute of Technology, Sweden.

My son David Iltis urged me to include the ways in which societies achieved buffers against nature, and my son John Iltis discussed the ramifications of the history of astronomy for the project. On trips across the country, my husband, Charles Sellers, offered numerous real-time examples of *natura naturans* and suggested the title *Autonomous Nature* as the best way of capturing the idea of the book. I am of course responsible for the results as well as any errors herein.

Carolyn Merchant
Berkeley, California

Introduction

Can Nature Be Controlled?

In the late fifteenth century B.C.E., a series of tsunamis radiated outward from a collapsed volcanic caldera in the Aegean Sea, just north of the island of Crete. Enormous waves rolled across the waters inundating land and peoples in their wake. In the path of the disaster was the Minoan civilization of Crete. It is hypothesized that this calamity, caused by a volcanic eruption, may have destroyed the ancient center of Minoan power, its natural resources, naval dominance, and culture.[1]

What are the implications of such chaotic events for science, politics, and the human future? Is the natural world fundamentally unpredictable, and, if so, can humans learn to live within it? Can science predict and therefore control the outcomes of catastrophic events? The historical relationship between the unpredictability of nature and human efforts to control it is the subject of this book. In it, I look at meanings of "nature" and the "natural" and at the roots of ways to manage the natural world. I argue that twenty-first century humanity is in the throes of a paradigm shift, one that is triggered by two factors: the rise of the new sciences of chaos and complexity and by climate change as the most widespread catastrophe for the human future. Chaos and complexity, unlike the more mechanistic sciences of the past, challenge our ability to predict with perfect certainty. Because chaotic systems are sensitive to initial conditions, uncertainty increases exponentially with elapsed time. Complexity, which deals with an extremely large number of dynamic sets of relationships, limits the degree of predictability. Climate change is both global in scope and cumulative in effect, reflecting these uncertainties and limits to predictability. New ways of living within the everyday world are therefore needed.

In what follows, I explore the prehistory of chaos and complexity theory from ancient times in the Western world to the (so-named) Scientific Revolution of the seventeenth century. I look at ideas of an unpredictable, unruly, and recalcitrant nature that captured the imaginations of ancient, medieval, and early modern European peoples and triggered efforts by seventeenth- and eighteenth-century scientists to find ways of predicting and controlling the world around them. In so doing, I hope to cast a new light on ways of thinking about nature and science in the past.

2   *Introduction*

A New Paradigm for the Twenty-First Century

In 1962, Thomas Kuhn published his foundational work, *The Structure of Scientific Revolutions.*[2] In it, he argued that major transformations in the history of science have been triggered by anomalies that do not fit into accepted theories. These transformative moments resulted in new scientific paradigms that over time became accepted as mainstream science. Such changes, however, can reach far beyond the confines of the scientific laboratory or institution to encompass society, politics, philosophy, and ethics. The Scientific Revolution of the seventeenth century is a prime example of a transformation in science that wrought changes in the wider society. Its greatest achievement was its understanding of the earth as a mechanism that could be understood, predicted, and controlled. A second example is the Scientific Revolution in Relativity Theory and Quantum Mechanics in the early twentieth century that followed the great triumphs of classical Newtonian mechanics in hydrodynamics, thermodynamics, and electromagnetism and in mathematics by differential equations, probability theory, and statistics. These advances, however, began to introduce uncertainties into the mechanistic worldview of Newtonian science.

Today, in twenty-first century America, we are in the midst of a third major paradigm shift. That paradigm is the wide-reaching societal transformation triggered by the rise of chaos and complexity science in the 1970s and 1980s.[3] The main achievement of seventeenth-century science was predictability; the main feature of the chaos paradigm is unpredictability. Although some scholars prefer to emphasize historical continuities and the history of science as prediction, in this book, I focus instead on transformative periods and the history of unpredictability.

Before proceeding further, however, a word of caution is needed. Although predictability and unpredictability are often cast as polar opposites, it should be noted that a range of possibilities exists between them, as well as levels and degrees of chaos, and a mixing of chaotic and deterministic systems. Charlotte Werndl points out that "it is widely believed and claimed by philosophers, mathematicians, and physicists alike that chaos has a new implication for unpredictability, meaning that chaotic systems are unpredictable in a way that other deterministic systems are not. . . . " But, she notes, there can be combinations of order and disorder at both large and small scales from the everyday world of macro-predictability to the subatomic world of micro-unpredictability, as well as predictable patterns in chaotic systems. Thus while one of great achievements of the Scientific Revolution of the seventeenth century was predictability in terrestrial and celestial mechanics, with wide-ranging consequences for the world in which we live today, subsequent advances in probability theory, statistics, quantum mechanics, relativity, chaos theory, and quantum computing have introduced uncertainties, variabilities, and probabilities into all fields of science.[4]

A major reason that I believe a history of unpredictability is important is the role that natural disasters and climate change play in the twenty-first century. Unpredictability is especially apparent in events such as hurricanes, tornadoes, earthquakes, droughts, and floods. And although climate change is largely anthropogenically caused and scientists can model and predict much of its extent, many outcomes are the result of unanticipated tipping points and complex cascading effects that, even if predicted, cannot be controlled. Because of the complexities of oceanic and atmospheric interactions, the moments, locations, results, and long-term effects of weather and climate make decision making highly problematic. The skepticism of climate-change doubters and the seeming inability of some lawmakers to promote the technologies and develop the regulations to manage and control changes and to prevent them from affecting individuals, communities, markets, and governments is of paramount concern. The long-term future of the earth itself is at stake.

In her book, *Complexity and Sustainability* (2013), Jennifer Wells argues that complexity represents a comprehensive new framework that encompasses the physical, biological, and social sciences. The transition to chaos and complexity that began in the 1970s and 1980s in the natural sciences now includes social theory, philosophy, and ethics and applies to the future of life on the planet. Wells writes that

> visions of complexity and sustainability have been emerging and developing in consort. Complexity theories have helped to rid us of falsely assuring assumptions of control and stability, and rather, show just how our planet is uncertain, vulnerable and resilient. . . . As such, complexity also touches the core of some of our oldest philosophical questions . . . all of which are necessary to understanding our current moment of momentous global change.[5]

Sustainability represents a new goal toward which human efforts must now be directed. New policies and ethics that deal with the environmental crisis and especially with global climate change are vital to new ways of living within a nature now characterized by uncertainties, nonlinearities, and unforeseen events. "In just a few decades," Wells argues,

> complexity science has flourished into not just a major new field, but also *the basis of a new paradigm*. Discoveries of complex dynamic systems, relevant across the widest range of area of reality—physical, living, and social systems—may play a central role in rearticulating our goals of sustainability.[6]

The big advantage of the clocklike, mechanistic universe was that it dealt with closed systems isolated from the environment. From the knowledge of initial conditions, the state of a particular system for any time in the future could be predicted with great accuracy. Living within a mechanistic world in which predictions hold gives confidence in everyday life that crossing bridges, flying across oceans, and lighting and heating our homes will be safe and secure. Only in very rare instances does chance disrupt such expectations—an earthquake of extremely high magnitude causing a bridge to collapse, a sudden lightning strike of such force that a plane is broken into pieces, a tornado that sweeps up a home

4   *Introduction*

and destroys it. But such external environmental disruptions are the result of unpredictable random events in situations that are highly localized in time and place. Within the mechanistic paradigm they are extremely unusual, not the usual events of everyday life. In the chaotic paradigm, on the other hand, external, environmental factors can play a major role, becoming the usual, rather than the unusual. Unpredictability and limited predictability become the new norms.

Meanings and Levels of Chaos

Biologist Daniel Botkin, in *The Moon in the Nautilus Shell* (2012) argues that there are several levels of chaos and hence of unpredictability, from well-behaved chance all the way to complete chaos without form or structure (in the sense of the ancient Greeks and Romans). The most predictable form follows probability theory, an example being the throwing of dice. The outcome over many, many throws follows a pattern of inherent randomness with the most likely outcome being seven. "While each throw has an unknown outcome, overall the game has a very regular, reliable, predictable behavior." Over time, the probabilities are well ordered and dependent only on the dice without the external environment playing a role. "This," says Botkin, "is the most predictable kind of chance: a process that has some inherent randomness but whose probabilities—chances of what will happen—are fixed forever and independent of everything else." Probability theories have their roots in eighteenth- and nineteenth-century advances by mathematicians such as Jacob Bernoulli (1654–1705), Carl Friedrich Gauss (1777–1855), and Pierre Simon LaPlace (1749–1827).[7]

The next level of chaos includes the environment. Here predictability depends on knowledge of the interactions among the parts of the system being studied and external environmental factors, such as the behavior of a pack of wolves in relation to conditions such as snow and the availability of food. The better the connections are understood, the better the ability to predict on average what the pack might do. "But once again," Botkin points out, "we cannot know perfectly, exactly, the future of any specific series of events." Probability and statistics are of special interest in applications in forestry and environmental science. These mathematical tools are especially useful in dealing with localized environmental problems such as the impact of fire or disease on forests or the role played by the introduction or loss of a species in a grassland, lake, or river system.[8]

The next levels of uncertainty introduce ever increasing degrees of randomness. Here probabilities and hence predictions can be calculated with rapidly diminishing reliability. "And so on downward," Botkin continues, "until one reaches what I will call complete chaos, an imaginary world where there is no cause and effect and that world is therefore without form, structure, and anything that we would call 'understandable' in normal terms."[9] At these higher levels of uncertainty, new ways of understanding and living within nature become important. Climate change is the most challenging aspect of uncertainty in the twenty-first century because of its global dimensions.

Botkin argues, however, that computers give us an advantage in living in this new world of uncertainty. We do not dwell in a world of complete chaos of the type envisaged by the Greeks, but one of multi-leveled complexity. Because computers can handle vast amounts of data with very high speeds, they go beyond the limitations of older models and methods based solely on probability theories and stochastic processes. Computers allow us to make predictions that involve chance and to adapt to a world in which internal and external factors, living and nonliving things, and mechanistic and organic worldviews are blurred and blended.[10]

While Botkin maintains an agnostic position on the human role in climate change, he nevertheless advocates real-world solutions to confront numerous, very real environmental problems. "Whether or not we cause climate change, these changes greatly affect us and all life, and we need to take these changes seriously while remaining scientifically objective." Actions include reversing the effects of deforestation, overfishing, species depletion, invasive species, environmental pollution, and threats to biodiversity. He further notes that in contrast to received views, the greatest government subsidies are actually for fossil fuels and nuclear power, while those for solar and wind energy, by contrast, receive a mere pittance.[11]

What we need, argues Botkin, is "a fundamental change in our paradigm," one in which we understand that "complicated, intricate, always changing system[s] . . . respond to novel input[s]." We must therefore give up the comfort and security of the clocklike world and embrace the complexities of nature. "We who work on environmental sciences and on global warming need to open ourselves to a much greater variety of ways of thinking about nature. We need to develop forecasting methods appropriate for always-changing, non-steady-state systems where chance—randomness—is inherent." Acknowledging that we live in a world in which chance and randomness are the norm could lead to new ways of engaging with nature.

> [W]e must accept nature for what we are able to observe it to be, not for what we might wish it to be. Accepting this perception of nature, we discover that we have the tools to deal with it. And once we realize we have the tools, this new idea of nature takes on its own appeal.[12]

What Botkin suggests is that we need to accept the fact that we are now living in a world defined in terms of several levels of chaos, in which there are levels of predictability, and in which the random actions of nature are the new norm. We therefore need new policies, new ethics, new scientific tools, and new ways of living within this new nature.

Wells's and Botkin's points about the importance of chaos and complexity are echoed by Sandra Mitchell in her book *Unsimple Truths: Science, Complexity, and Policy*. Mitchell, in a section on "Shifting Paradigms in Epistemology," notes that

> The successes of the Scientific Revolution of the seventeenth century in providing simplifying, unifying representations, in particular Newton's laws of

6   *Introduction*

> motion and his law of universal gravitation, led philosophers to define what they would admit as reliable knowledge in like terms. . . . But the world of Newtonian science did not persist. Twentieth century physics challenged some of its most fundamental assumptions.

In making her "Case for Complexity," Mitchell argues that complexity requires new modes of understanding. In particular, "we need to expand our conceptual frameworks to accommodate contingency, dynamic robustness, and deep uncertainty."[13]

While Wells, Botkin, and Mitchell make the case that a new paradigm and new ways of understanding and describing the world of the twenty-first century are needed, my book looks at the prehistory of chaos and complexity from ancient times to the Scientific Revolution of the seventeenth century. I do not claim to make a comprehensive history of that era, but rather to look specifically at authors who dealt with unpredictability and chaos. Nor do I attempt to write a history of challenges to Newtonianism and mechanistic thought in the eighteenth through twentieth centuries. Instead, although I recognize the advances that took place in the sciences of hydrodynamics, thermodynamics, and electromagnetism, in the mathematics of probability and statistics, and in the revolutions in relativity theory and quantum mechanics (see below and Chapters 6 and 7), my goal here is to look at early thinkers in the Western world to see how they met the challenge of dealing with uncertainty, chaos, and unpredictability.

Climate Change

The world of the twenty-first century is one in which chance and randomness constitute our new reality. The new paradigm is exemplified most dramatically by global climate change. In 2013, observations on the level of atmospheric CO_2 conducted over several decades at Mauna Loa Observatory in Hawaii hit a landmark 400 parts per million (ppm), reaching 403 ppm in June 2015. Since the Industrial Revolution of the eighteenth century and the burning of fossil fuels—an era named the anthropocene—levels of CO_2 have risen dramatically. Many believe that although a safe level of CO_2 is 350 ppm, that level could rise to a dangerous 450 ppm within a few decades. Such rises in CO_2 could trigger an ecological tipping point in which a tiny fluctuation could initiate a series of cascading effects on organisms and their interactions and dependencies. Such a point of no return would herald a series of drastic impacts on human life support systems. At such threshold points, the changes and effects become fundamentally unpredictable.[14]

According to Edward Lorenz, chaotic behavior can be observed in many natural systems, especially weather and climate. In such dynamical systems, small differences in initial conditions can result in widely differing outcomes, making long-term predictions impractical if not impossible. While such systems might, in principle, be deterministic, over time random unpredicted changes can occur. In climate change, for example, small changes in global temperature can lead to unexpected and abrupt changes in ecosystems from oceans to rainforests. Here the rapid doubling of errors precludes great accuracy in real world forecasting. Lorenz argued that we cannot predict the results of small effects such as a butterfly on the weather, because the atmosphere is unstable with respect to perturbations of small amplitude. We don't know how many small effects there are (such as butterflies) or even where they are located. We can't even set up a controlled experiment to find out if the atmosphere is unstable because we can never know what might have happened if we hadn't disturbed it. Unstable dynamic systems, such as weather and climate, behave differently than stable systems, such as the planetary systems described by Isaac Newton. In such cases predictability is limited or unfeasible.[15]

The problem of predictability and control was pushed further by Ilya Prigogine in his work with Isabelle Stengers on *Order out of Chaos*. While classical thermodynamics, discovered in the nineteenth century, described equilibrium and near-equilibrium situations, as in the steam engine and refrigerator, in Prigogine's far-from-equilibrium thermodynamics, as found in hydrodynamics, many chemical processes, and evolution, a new reorganization can occur in which order can emerge out of chaos. In such situations irreversibility and nonlinearity can lead to self-organization. Irreversibility and nonlinearity increase the role of fluctuations and lead to bifurcations (divisions) in which the system can go in several directions because nonlinear equations can have several different solutions. The outcome, therefore, cannot be predicted with certainty and the control of nature is problematic.[16]

As Prigogine puts it:

> The important element is that unstable systems are not controllable. . . . The classical view on the laws of nature, on our relation with nature, was domination. That we can control everything. If we change our initial conditions, the trajectories slightly change. . . . But that is not the general situation. . . . We see in nature the appearance of spontaneous processes which we cannot control in the strict sense in which it was imagined to be possible in classical mechanics. . . . The world in which we are living is highly unstable. . . . [17]

Today we are living within a new paradigm in which nature is autonomous. Our world is characterized, not by determinism and predictability, but by several levels of unpredictability and degrees of forecasting. During the twenty-first century we have the opportunity to halt and even reverse the mounting effects of global warming on human and nonhuman life. As I wrote in *Earthcare* and other works in the 1990s, chaos theory

> reinforces the idea that predictability while still useful, is more limited than previously assumed and that nature, while a human construct and a representation, is also a real, material, autonomous agent. . . . Because nature is fundamentally chaotic, it must be respected and related to as an active partner through a partnership ethic.[18]

We must find new ways to live within nature, to stem the effects of climate change, and to halt the loss of species and collapse of populations. Ways of living

8   *Introduction*

within the new chaos/complexity paradigm will engage the best scientific, engineering, political, and ethical practices we can develop. Nature, experienced as an active, creative, often uncontrollable force, is the new norm, one that will challenge our own creativity, imagination, and vision. "This disorderly, ordered world of nonhuman nature," I wrote, "must be acknowledged as a free autonomous actor, just as humans are free autonomous agents." Moreover, "Humans as the bearers of ethics would acknowledge nonhuman nature as an autonomous actor that cannot be predicted or controlled except in very limited domains."[19]

Meanings of Nature

How did past thinkers use the term "nature" and which of those meanings are most relevant to this book? Raymond Williams has argued that the word "nature" is the most complex term in the English language. First of all, it refers to the entire physical universe, or cosmos, including its atoms, elements, molecules, compounds, and the constituent parts of the animate and inanimate worlds. It also includes physical changes, such as those produced by mechanical, hydrodynamic, thermodynamic, and electromagnetic forces and processes. Second, it designates living organisms and their interactions with their physical surroundings and the ways in which components pass across boundaries and through ecosystems. Third, it encompasses humans and human activities that engage with and alter the animate and inanimate worlds. Fourth, it includes human social, political, and cultural institutions as well as an individual entity's character or qualities.[20]

Throughout history, nature and parts of nature have been personified both as male and female. In Egypt, the male god, Geb, symbolized the earth; in Greece, Orpheus was a charmer of living and nonliving things; in Rome, Pan was the god of the forest. But for the most part, Nature was female. "She" was portrayed as a goddess, virgin, mother, and witch. *She* acted both on her own and under the direction of the gods/God. She was kind, benevolent, and caring, but could also be wild, unruly, and vindictive.

In this book, I explore the meanings associated with the Greek and Roman words for nature—*physis* and *natura*—and the evolution and meanings of the terms *natura naturans* (nature creating, evolving, and changing) and *natura naturata* (nature as experienced in the everyday world) from ancient times through the Scientific Revolution. These words have gendered meanings and histories that played a role in the thought of many, although by no means all, natural philosophers. The Greek noun *physis* and the Latin noun *natura* are both feminine and throughout history have had (so-called) female characteristics associated with them. Interesting, as well, are the gendered implications of the modifiers *naturans* and *naturata*. Both stem from the Latin verb of the first conjugation: *naturo, naturare, naturavi, naturatus*. *Naturans* is the present participle (equivalent to adding "ing" in English), meaning the activity of creating or naturing, i.e., nature in the active sense or nature doing what nature does. *Naturata* is the feminine past participle, designating that which has already been created or natured; the passive result of the causal series. Of the two participles, *naturans* was associated with an active creator and a sometimes rebellious nature, while *naturata* was associated with the female nature of the created world.[21]

The meanings and implications of the noun *Natura* (nature) and the participles *naturans* (creating, evolving) and *naturata* (the everyday, created world) reflect Greek frameworks that stem from the philosophies of Parmenides, Heraclitus, Plato, and Aristotle as carried down through Arabic translations to the Christian Middle Ages, Renaissance, and Scientific Revolution. From the dilemma of Parmenides's unchanging Being (or all is Substance) versus Heraclitus's all is Change (and no Substance), Plato and Aristotle introduced the concepts of Form and Matter. For Plato, the two were separate entities, whereas for Aristotle they were combined in individual things. But Form was the ideal, unchanging pattern, Matter the changing substrate. Form was masculine, matter feminine (see Chapter 1).

Biologist Daniel Simberloff, in a paper on the "Succession of Paradigms in Ecology," argues that until the late nineteenth century, "the goal of philosophy and science" was "to understand the ideal [F]orms." Plato's Forms were ideal types that when imbedded in matter introduced variations, producing the world of appearances. The Scientific Revolution of the seventeenth century witnessed the triumph of Platonic Form, exemplified by deterministic mechanics, and with it predictability. Newton's ideal bodies followed ideal trajectories in abstract space. Actual material bodies (such as planets and projectiles) deviated only slightly from the ideal patterns described by the abstract symbols and numbers of the calculus. Here the Forms do not change, only the material content changes.[22]

What led to the breakdown of the dominance of ideal Forms was the emergence of probability theory, statistics, and stochastics in mathematics and the acceptance of randomness and "noise" in the material world. Here the domination of Form began to give way to the predominance of matter and process. Probability theory has its origins in the need to predict risks in insuring vessels at sea and the uncertainties of evidence in court cases. Attempts to predict outcomes, using games of chance as a model, were undertaken by Pierre de Fermat, Blaise Pascal, and Christian Huygens in the seventeenth century. In the eighteenth and nineteenth centuries, probability theory was refined by Jacob Bernoulli in his law of large numbers, by Laplace in his *Analytic Theory of Probabilities* (1812), and applied to astronomy by Carl Friedrich Gauss using his theory of least squares. In the late nineteenth century, statistical mechanics developed by Ludwig Boltzmann, James Clerk Maxwell, and J. Willard Gibbs helped to explain the random motions of particles in gases under variations in temperature. Probability theory and statistics were applied in both the sciences and the social sciences as well as health and medicine.[23]

By the late nineteenth century, a materialist, process-oriented perspective began to supersede that of ideal forms and unchanging types. Simberloff puts it this way:

> The nineteenth century was dominated by a deterministic mechanics, hypostatized by twin hypothetical ideal beings, Laplace's Demon and Maxwell's Demon. The former could predict in Newtonian, cause-and-effect, action-reaction fashion the complete state of the universe, given knowledge

10   *Introduction*

of the positions and velocities of all its particles for a single instant. The latter could violate the second law of thermodynamics, and in so doing, construct a perpetual motion machine. . . . [T]he beginning of the revolution against determinism [begins] with Maxwell's observation that the velocities of gas particles are [actually] distributed according to a statistical law.[24]

Darwinian evolution and the recognition by Gregor Mendel that "both similarity and variation are produced by the same mechanism" was a further manifestation of the breakdown in ideal forms and types. The world was an evolving, changing, developing place and ongoing change came from within matter, as opposed to a world determined by a transcendent deity or a pure Form.[25]

The deterioration of determinism continued during the second Scientific Revolution with Einstein's Theory of Special Relativity (1905), Plank's Quantum Theory (1900), and Heisenberg's Uncertainty Principle (1927). These early twentieth-century revolutions introduced a new paradigm of uncertainty at the highest velocities and smallest dimensions, with implications in the everyday world for nuclear reactors, lasers, transistors, and LED (light-emitting diode) lights.

Then, in the mid-twentieth century, ecology, based on open (rather than closed) systems, in which matter and energy move across boundaries, underwent a similar revolution away from ideal types and deterministic predictions.[26] The chaos and complexity revolutions of the late twentieth century carried the level of uncertainty and unpredictability into the world of daily life as manifested most dramatically in unpredictable weather patterns, the tipping points and cascading effects of climate change, and far-from-equilibrium (order-out-of-chaos) thermodynamics.

In his introduction to *Order out of Chaos* by Ilya Prigogine and Isabelle Stengers, Alvin Toffler writes:

[W]hile some parts of the universe may operate like machines, these are closed systems, and closed systems, at best, form only a small part of the physical universe. Most phenomena of interest to us are, in fact, *open* systems, exchanging energy or matter (and, one might add information) with their environment. . . . This suggests, moreover, that most of reality, instead of being orderly, stable and equilibrial, is seething and bubbling with change, disorder, and process.[27]

And Richard Kautz in *Chaos: The Science of Predictable Random Motion* (2011) states:

For three centuries the laws of dynamics were successfully applied to system after system, until it was believed that the future was entirely predictable, if we only knew all the forces involved. This picture of a clockwork universe began to lose credibility first in the atomic domain with the advent of quantum mechanics during the 1920s. At that time, physicists discovered that microscopic systems are irreducibly random and unpredictable. With the discovery of chaos in the 1960s, however, unpredictability was extended to the macroscopic world. . . . Chaos is a strange new effect that forces us to revise downward our estimate of what science can do. Scientists now realize that when chaos is present their knowledge of the future may be extremely limited even in the macroscopic world.[28]

Mechanism continues, however, in the everyday world of engineered bridges, skyscrapers, and airplanes in which we spend many of our waking moments. And mechanistic explanations are still the norm in molecular biology, genetics, neuroscience, parts of biochemistry, chemistry and physics, and parts of engineering and biophysics. But although Newtonian mechanics still describes the world in which we live and move, that world is a limited domain of the far larger, unpredictable, and dominant domain now characterized by chaos and complexity science. It is that world to which twenty-first century life must adapt (see Table 1.0).

Natural Disasters

Throughout this book, I provide vignettes of sudden, unpredicted, chaotic events and people's responses to them. Such dramas include, e.g., the eruption of Mt. Vesuvius volcano in 79 C.E., the spread of the Bubonic Plague in 1348, and the Lisbon Earthquake of 1755. I ask to what extent were they acts of autonomous nature, as well as socially and culturally constructed "nature-induced disasters"? "Over the past two decades," writes Christof Mauch, "scholars have increasingly come to accept that natural catastrophes are never 'natural' in the true sense of the word; instead, they should be understood as both physical events and social or cultural occurrences." Christian Pfister clarifies that "a more precise alternative would be the phrase *nature-induced disaster*, which reflects the fact that such catastrophes are brought about by natural phenomena without obscuring their anthropogenic dimensions." Such "disasters are commonly understood as unpredictable outbreaks of elemental forces that have a sudden, destructive impact on human affairs." These include, for example, "'rapid-onset hazards' such as frosts, floods, storm tides, windstorms, volcanic eruptions, rockslides, earthquakes, and tsunamis, the impacts of which are magnified by their unpredictability."[29]

"Natural disasters" therefore have both physical and human dimensions. Geomorphologic and tectonic processes are operative in the case of earthquakes, volcanoes, and tsunamis; weather and climate are implicated in droughts, freezes, floods, and hurricanes; and plagues are transmitted by bacteria, viruses, rodents, and other biological vectors. But "disasters" are also social constructions, associated with human habitats, destruction, and death. They are magnified by urban crowding, collapse of buildings, destruction of property, and widespread epidemics and are made all the more tragic by conditions of poverty, race, class, and gender. Were they to occur in remote, uninhabited areas they might not even be called disasters. To the extent that such geological and human factors can be identified

Table 1.0 Meanings of Nature

Period	Philosophers	Nature	Cosmic Nature	Earthly Nature	Major Concepts	Related Concepts	Chaotic Nature
Ancient Greece	Hesiod, Parmenides, Heraclitus, Plato, Aristotle	*Khaos* (chaos) *Phusis/physis* Nomos (law)	Stellar Sphere, (*De Caelo*)	Earth, air, fire, and water	Being, Change; Form, Matter	Qualities, Quantities; Essences; *Telos* (Goal)	Destruction of Crete, ca. 1480 B.C.E.
Ancient Rome	Lucretius, Plotinus	*Natura Creatrix*	Fixed Stars, Central Earth	Atoms	One (*Monad*), Mind (*Nous*), Soul (*Anima*)	Swerve, Rebellious Audacity	Eruption of Mt. Vesuvius, 79 C.E.
Middle Ages	Augustine, Erigena, Aquinas, Bonaventure	*Natura creans/ naturans, Natura creata/ naturata*	Empyrean Heaven, God	World Soul, Spirit, Body	Body, Soul, Spirit	Salvation, Resurrection	1348 Earthquake, Italy
Renaissance	Nicholas of Cusa, Alberti, Da Vinci, Bruno	*Natura naturans/ Natura naturata*	Closed Cosmos	World Soul, Spirit, Body	Body, Soul, Spirit	Life after Death	Black Death, outbreaks 1348–1374
Scientific Revolution	Bacon, Boyle Spinoza Newton, Leibniz	Laws of nature (mathematics) Experimentation Natural law	Infinite Universe	Corpuscles, particles, atoms	Matter in motion	Quantities, Gravity, Machines	Lisbon Earthquake, 1755
21st Century	Lorenz, Prigogine	Chaos, Complexity	Multiple Universes	Quarks, Quanta	Digital	Computers, Biotechnology	Climate Change

as possibilities, such catastrophes are "disasters" waiting to happen. Nevertheless disasters have different levels of predictability. This book deals with the history of efforts to understand, predict, and control autonomous nature and asks, to what extent is it possible for science to predict chaotic events?[30]

In looking at the roots of efforts to predict nature, I likewise examine the material conditions that increasingly allowed for the possibility of controlling nature, such as agricultural surpluses, water management, the rise of cities, trade, and so on—conditions that created a buffer for humanity against unruly nature. In the epilogue, I bring the argument forward in time by looking at the implications of these concepts for the idea of nature in the present era and why it is critical to think about a new chaotic paradigm for the twenty-first century and ways to respond to it culturally, politically, and ethically.

Argument of the Book

Part I on "Autonomous Nature," examines the ways in which nature was described in the Greco-Roman world, the Middle Ages, and the Renaissance, giving vivid examples of disasters that depict a chaotic nature gripping the minds of those living at the time. In Chapter 1, I examine the use of the terms *Khaos* (chaos), *Phusis* (*Physis*), *Natura* (nature), and *Nomos* (law) by Greeks and Romans to depict nature as disorderly/unlawful and as orderly/lawful. Chapter 2, "Christianity and Nature," looks at the ways in which Christian theologians in the Middle Ages, such as Augustine, Erigena, Aquinas, and Bonaventure, merged ideas of nature from the Greco-Roman worlds with the God of Christianity. These theorists developed a framework in which nature was seen both as an active creative force (*natura creans*, or *naturans*) (God) and as the created world (*natura creata*, or *naturata*). Chapter 3, "Nature Personified: Renaissance Ideas of Nature," examines how a personified Nature acting in accordance with God's plan was characterized not only by a rational, logical, creating nature, but a nature that might also become willful and unruly. The tensions between these two ideas appeared in late medieval and Renaissance philosophy, art, and literature.

In Part II on "Controlling Nature," I investigate breakthroughs in experimentation and mathematics that gave scientists the tools and technologies to understand, predict, and manage a nature now described by some as a lifeless machine rather than a willful witch. During the Scientific Revolution of the seventeenth century, the goal of controlling a free, recalcitrant nature through the laws of nature and natural law became paramount. Chapter 4, "Vexing Nature" examines the role of Francis Bacon in the origins of the experimental method. Through the "vexations of art," or the use of technologies to constrain and examine nature, it could be studied and improved. Through experimentation as an intervention into nature, it could be manipulated for the benefit of humankind. In Chapter 5 on "Natural Law," I look at the ways in which Baruch Spinoza developed the ideas of *Natura naturans* and *Natura naturata* into a rational framework. The universe itself was a logical coherent whole, the expression of which emerged by necessity from the laws of logic. Chapter 6 on the "Laws of Nature" expands the argument

14   Introduction

for mathematics and rationality to include the natural philosophies of Isaac Newton and Gottfried Wilhelm Leibniz. Euclidean geometry, analytic geometry, and calculus gave mathematicians new tools with which to understand and predict the motions of celestial and terrestrial bodies. But the question of God's role in a world in which comets might disrupt the solar system and the problem of disorder in the everyday world continued to engage Newtonians and Leibnizians alike.

The Epilogue, "Rambunctious Nature in the Twenty-First Century," looks at the implications of chaos and complexity as a new paradigm for the future. Here I discuss ways in which mechanistic science was transformed by the rise of probability theory and statistics in the nineteenth century, the revolution in relativity and quantum mechanics in the early twentieth century, and by chaos and complexity in the late twentieth century. I then bring the problem of an unpredictable and potentially uncontrollable nature forward to the twenty-first century. I suggest that *the earth as we know it today* might be very different in the future (e.g., through nuclear apocalypse, climate change, genetically engineered species, and so on).[31] The tensions between *natura creans/naturans* and *natura creata/naturata* have reappeared as a new paradigm characterized by unpredictability—in the sciences by chaos and complexity theory and in society by climate change. New posthuman, postwild, and postmodern concepts of nature and the interlinked relationships between nature and culture call for a new ethic for the twenty-first century, one I call partnership ethics.[32]

Autonomous Nature focuses on the history and gendered implications of nature as an independent force and on two concepts, nature naturing and natured nature—nature as a creating, evolving force and nature as the created world of everyday life. Why should we care about the meaning of past terms such as nature creating and nature created? Because today the idea of autonomous nature includes those meanings, but now encompasses geological, evolutionary, and quantum-mechanical forces along with human impacts on the earth's species, soils, water, atmosphere, and climate. *And* because past meanings of *natura naturans* and *natura naturata* help to illuminate the prehistory of one of the most significant changes in the history of science and indeed the history of humankind—the role of chaos and complexity theories as a new paradigm for the twenty-first century.

Notes

1 On the eruption of the Aegean volcano, Santorini, the tsunamis generated from its collapsed caldera, and its role in the destruction of Minoan civilization on the Greek island of Crete, see http://www.drgeorgepc.com/AtlantisDestruction.html and http://en.wikipedia.org/wiki/Santorini.

2 Thomas Kuhn, *The Structure of Scientific Revolutions* (Chicago: University of Chicago Press, 1962); Peter Harrison, "Was There a Scientific Revolution?" European Review 15 (2007).

3 Edward Lorenz, *The Essence of Chaos* (Seattle, WA: University of Washington Press, 1993); James Gleick, *Chaos: Making a New Science* (New York: Viking, 1987); M. Mitchell Waldrop, *Complexity: The Emerging Science at the Edge of Order and Chaos* (New York: Simon and Schuster, 1992); Stephen Kellert, *In the Wake of Chaos: Unpredictable Order in Dynamical Systems* (Chicago: University of Chicago Press, 1992).

Can Nature Be Controlled? 15

4 Charlotte Werndl, "What Are the New Implications of Chaos for Unpredictability?" *British Journal for the Philosophy of Science* 60 (2009): 195–220, see esp. pp. 196–197, 211–212, 215, quotation on p. 195.

5 Jennifer Wells, *Complexity and Sustainability* (New York: Routledge, 2013), pp. 1–18, quotation on p. 17.

6 Wells, *Complexity and Sustainability*, pp. 52–87, 284–297, quotation on p. 87, italics added.

7 Daniel Botkin, *The Moon in the Nautilus Shell: Discordant Harmonies Reconsidered, From Climate Change to Species Extinction, How Life Persists in an Ever-Changing World* (New York: Oxford University Press, 2012), pp. 166–180, quotations on p. 173.

8 Botkin, *Moon in the Nautilus Shell*, quotation on p. 174; Stacey J. Schaeffer and Louis Theodore, *Probability and Statistics Applications for Environmental Science* (Boca Raton, FL: CRC Press, 2007).

9 Botkin, *Moon in the Nautilus Shell*, quotation on p. 174.

10 Botkin, *Moon in the Nautilus Shell*, Ch. 8, "The Forest in the Computer: New Metaphors for Nature," pp. 157–180, esp. 179–180.

11 Botkin, *Moon in the Nautilus Shell*, quotation on p. 286; Botkin, "Carbon Dioxide and Temperature: Who Has Led Whom?" http://www.danielbbotkin.com/2013/03/04/carbon-dioxide-and-temperature-who-has-led-whom/; Botkin, "It's not Solar and Wind, but Fossil Fuels, Nuclear and Biofuels that Reap the Most Subsidy Benefits," http://www.danielbbotkin.com/2013/08/15/its-not-solar-and-wind-but-fossil-fuels-nuclear-and-biofuels-that-reap-the-most-subsidy-benefits/.

12 Botkin, *Moon in the Nautilus Shell*, pp. 285–286, quotations on pp. 286, 285, 179.

13 Sandra Mitchell, *Unsimple Truths: Science, Complexity, and Policy* (Chicago: University of Chicago Press, 2009), pp. 5–19, quotations on pp. 11, 12, 5.

14 Justin Gilless, *New York Times*, May 13, 2013. For June 2015 and weekly trends, see, http://www.esrl.noaa.gov/gmd/ccgg/trends/weekly.html. On the continuous rise in global warming without what was previously thought to be a "temperature change hiatus," during the early 2000s, see http://www.nature.com/news/climate-change-hiatus-disappears-with-new-data-1.17700. For background on the Mauna Loa readings see Botkin, *Moon in the Nautilus Shell*, Ch. 12, "The Winds of Mauna Loa: Climate in a Changing World," pp. 245–287; on James Hansen's concept of tipping points, see p. 280. On "Earth in the Anthropocene," and the concepts of threshold, tipping points, and abrupt change see Jennifer Wells, *Complexity and Sustainability* (New York: Routledge, 2013), Ch. 8, pp. 212–231, esp. 225 and Ch. 9, "Complexity and Climate Change," pp. 232–266.

15 Edward N. Lorenz, "Predictability: Does the Flap of a Butterfly's Wings in Brazil Set off a Tornado in Texas?" (1972) reprinted in Lorenz, *Essence of Chaos*, pp. 181–184; Edward N. Lorenz, "Crafoord Prize Lecture," *Tellus* (1984): 36A, 98–110; Carolyn Merchant, *Key Concepts in Critical Theory: Ecology*, 2nd ed. (Amherst, NY: Humanity Books, 2008), pp. 397–400 and "Introduction," by Merchant, taken from p. 34.

16 Ilya Prigogine and Isabelle Stengers, *Order out of Chaos: Man's New Dialogue with Nature* (New York: Bantam, 1984), p. 12–18. On Prigogine's work, see Erich Jantsch, *The Self-Organizing Universe* (New York: Pergamon Press, 1980).

17 Ilya Prigogine, "The Rediscovery of Time: Science in a World of Limited Predictability," paper presented to the International Congress on "Geist & Natur" [Spirit and Nature], Hanover, Germany, May 21–27, 1988, reprinted in Merchant, *Key Concepts in Critical Theory: Ecology*, quotation on p. 405 and "Introduction" by Merchant, taken from p. 35.

18 Carolyn Merchant, "Partnership Ethics: Earthcare for a New Millennium," in Merchant, *Earthcare: Women and the Environment* (New York: Routledge, 1996), quotations on p. 221.

19 Quotations are from Merchant, *Earthcare*, op cit., p. 221, Carolyn Merchant, "Reinventing Eden" in *Uncommon Ground: Rethinking the Human Place in Nature*, ed. William Cronon (New York: Norton, 1995; 2nd ed. 1996), p. 158, and Carolyn Merchant,

16　*Introduction*

Reinventing Eden; The Fate of Nature in Western Culture (New York: Routledge, 2003), p. 220 (2nd. ed. 2013, p. 186). On the question of whether nature is autonomous and if so how can it be recognized, respected, and restored, see the essays in *Recognizing the Autonomy of Nature: Theory and Practice*, ed. Thomas Heyd (New York: Columbia University Press, 2005), quotation from Merchant on humans as the bearers of ethics and nonhuman nature as an autonomous actor, on p. 8. In the Preface, Heyd writes,

> I was intrigued by the question whether nonhuman nature in its totality, or any part of it, may be said to *act* at all. If it reasonably could be said that it acts, even if only in an attenuated way, it would open up the space to ask whether this entailed moral responsibilities for human beings toward nature and perhaps for nature toward human beings." (p. ix)

In her essay, "Is Nature Autonomous?" Keekok Lee writes,

> The answer to the question posed by the title of this essay is "yes." . . . I argue that autonomy should be defined, on the one hand, negatively as what exists and continues to exist independently of human intentionality, control, manipulations, or intervention and, on the other hand positively in terms of being self-generating and self-sustaining and explain why the term may be applied to biotic as well as abiotic nature. I conclude by showing that autonomous nature is more than a social construct. (p. 54)

20　Raymond Williams, "Nature," in *Keywords*, 2nd ed. (New York: Oxford University Press, 1985 [1976]), pp. 219–224.

21　On the Latin variants of *natura*, *naturans*, and *naturata*, see http://www.democraticunderground.com/12182702. On the participles of the verb *naturo*, see http://pursuingtraditions.wordpress.com/category/natura-naturans-vs-natura-naturata/: "*Natura naturata* is a Latin term coined in the Middle Ages, mainly used by Baruch Spinoza meaning 'Nature natured', or 'Nature already created.' The term adds the suffix for the Latin feminine past participle (-*ata*) to the verb *naturo*, to create 'natured.'"

22　On the history of the change from Greco-Roman ideas of essences to probabilities, see Daniel Simberloff, "A Succession of Paradigms in Ecology: Essentialism to Materialism and Probabilism," *Synthese* 43, no. 1 (January 1980): 3–39, esp. pp. 4–5, quotation on p. 4.

23　Simberloff, pp. 9–10. On the history of probability and statistics, see Lorenz Krüger, Lorraine Daston, and Michael Heidelburger, eds., *The Probablistic Revolution* (Cambridge, MA: MIT Press, 1987), 2 vols.; Ian Hacking, *The Emergence of Probability: A Philosophical Study of Early Ideas about Probability, Induction, and Statistical Inference* (New York: Cambridge University Press, 2006); Stephen M. Stigler, *The History of Statistics: The Measurement of Uncertainty Before 1900* (Cambridge, MA: Harvard University Press, 1986); Laura J. Snyder, *Reforming Philosophy: A Victorian Debate on Science and Society* (Chicago: University of Chicago Press, 2006).

24　Simberloff, "Succession of Paradigms in Ecology," op. cit., p. 10.

25　Simberloff, "Succession of Paradigms in Ecology," op. cit., esp. 4–9, quotation on p. 7.

26　Simberloff, "Succession of Paradigms in Ecology," op. cit., pp. 10–11, 13–22. Simberloff's foundational paper was written in 1980, before the development and implications of chaos and complexity theories became significant.

27　Prigogine and Stengers, *Order out of Chaos*, "Introduction," by Alvin Toffler, p. xv.

28　Richard Kautz, *Chaos: The Science of Predictable Random Motion* (New York: Oxford University Press, 2011), p. 13.

29　Christof Mauch and Christian Pfister, eds., *Natural Disasters, Cultural Responses: Case Studies Toward a Global Environmental History* (Lanham, MD: Lexington Books, 2009), "Introduction," by Christof Mauch, pp. 1–16, quotation on p. 4; Christian Pfister, ibid., "Learning from Nature-Induced Disasters," pp. 17–40, quotations on p. 18.

30 On the social construction of natural disasters, see Charlotte Rossi, "How Are Natural Disasters Socially Constructed?" http://www.studymode.com/essays/How-Are-Natural-Disasters-Socially-Constructed-684217.html; Mark Pelling and Juha I. Uitto, "Small Island Developing States: Natural Disaster Vulnerability and Global Change," *Environmental Hazards* 3 (2001): 49–62; Chester Hartman and Gregory D. Squires, *There Is No Such Thing as a Natural Disaster: Race, Class, and Hurricane Katrina* (New York: Routledge, 2006); Terry Cannon, "Vulnerability, 'Innocent' Disasters, and the Imperative of Cultural Understanding," *Disaster Prevention and Management* 17, no. 3 (2008): 350–357.

31 Bill McKibben, *The End of Nature* (New York: Random House, 1989); Alan Weisman, *The World Without Us* (New York: Picador Press, 2007).

32 Bruno Latour, *We Have Never Been Modern* (Cambridge, MA: Harvard University Press, 1993); N. Katherine Hayles, *How We Became Post-Human: Virtual Bodies in Cybernetics, Literature, and Informatics* (Chicago: University of Chicago Press, 1999); Emma Marris, *Rambunctious Garden: Saving Nature in a Post-Wild World* (New York: Bloomsbury, 2011).

Part I

Autonomous Nature

1 Greco-Roman Concepts of Nature

On August 24, 79 C.E. Mount Vesuvius volcano, near Naples, Italy, exploded. "Broad sheets of fire and leaping flames blazed at several points, their bright glare emphasized by the darkness of night." Black clouds of ash, fumes, and pumice flew miles into the air for two days, raining down on the ancient city of Pompeii and burying it in debris. Pliny the Younger who at the age of eighteen survived the eruption wrote of its terror. His uncle Pliny the Elder perished. A few years later, the younger Pliny who had thrown off mounds of ash in his struggle for life summoned the courage to write about his uncle's experience:

> They debated whether to stay indoors or take their chance in the open, for the buildings were now shaking with violent shocks, and seemed to be swaying to and fro as if they were torn from their foundations. Outside, on the other hand, there was the danger of failing pumice stones, even though these were light and porous; . . .
>
> Elsewhere there was daylight by this time, but they were still in darkness, blacker and denser than any ordinary night, which they relieved by lighting torches and various kinds of lamp. My uncle decided to go down to the shore and investigate on the spot the possibility of any escape by sea, but he found the waves still wild and dangerous. A sheet was spread on the ground for him to lie down, and he repeatedly asked for cold water to drink.
>
> Then the flames and smell of sulphur which gave warning of the approaching fire drove the others to take flight and roused him to stand up. He stood leaning on two slaves and then suddenly collapsed, I imagine because the dense, fumes choked his breathing by blocking his windpipe which was constitutionally weak and narrow and often inflamed. When daylight returned on the 26th—two days after the last day he had been seen—his body was found intact and uninjured, still fully clothed and looking more like sleep than death.[1]

The experience of Pliny the Younger graphically illustrates the way in which peoples of the ancient world lived within autonomous nature. Volcanos, earthquakes, hurricanes, and disease, although not everyday events, loomed as exemplars of the wrath of the gods and of worldly chaos—unpredictable events over which humanity exercised no control.

Figure 1.1 Vesuvius in Eruption, with a View over the Islands in the Bay of Naples. Joseph Wright of Derby (1734–1797), ca. 1776–1780. Copyright Tate, London, 2015.

In this chapter, I discuss the Greco-Roman background to the idea of Nature as an active, creating entity versus Nature as the created world—ideas that during the Christian Middle Ages and Renaissance would become the concepts of *Natura creans/naturans* and *Natura creata/naturata*. The association of natural disasters with unpredictability and chaos has its roots in ancient philosophy. Especially important to later natural philosophy is the gendering of nature, matter, and chaos as female and Intellect, Ideas, and Form as male. I describe Plato's idea of the Demiurge as an intelligent craftsman who creates the visible world and compare it with the craftsman in the Hippocratic corpus who attempts to extract secrets from the body of nature and the human being. I also discuss Aristotle's merging of the creating and created within each individual object in the natural world and Lucretius's concept of *Natura Creatrix* as a creating deity. Both Lucretius and Plotinus portrayed Nature as lawful and orderly, but sometimes rebellious and disorderly, hence unpredictable and uncontrollable. These ideas would be elaborated in ensuing centuries as Christianity drew on and integrated Greek and Roman ideas with those of the *Bible* and in the concepts of experimentation and mathematization as ways of predicting and controlling autonomous nature developed during the Scientific Revolution.

Chaos

In his *Theogony* (ca. 700 B.C.E.), Hesiod depicts Chaos (*Khaos*, the gap) as the gaping void, or primordial condition out of which everything emerged. Chaos, as female progenitor of the gods, was the chasm or open gap that gave birth to the gods and the cosmos. Next came Gaia, the female earth, and Eros, or sexual desire.

> Verily at the first Chaos came to be, but next wide-bosomed Earth [Gaia], the ever-sure foundations of all the deathless ones who hold the peaks of snowy Olympus, and dim Tartarus in the depth of the wide-pathed Earth, and Eros (Love), fairest among the deathless gods, who unnerves the limbs and over-comes the mind and wise counsels of all gods and all men within them.[2]

For Hesiod, the Sky or Heaven (*Ouranos; Uranus*) was male and the Earth (*Gaia*) was female, a gendered division that was to play a central role in the history of science and the environment. "And Heaven came, bringing on night and longing for love, and he lay about Earth spreading himself full upon her."[3] The personifications of a male Sky and female Earth, along with Chaos (female and mother of the gods) as the gap between them, set the stage for later efforts to understand and control a chaotic, unpredictable world often depicted in female terms.

In attempting to understand the origins of nature, its laws and activities, and how to predict what a chaotic Nature might do next, Greek and Roman philosophers devised a number of theoretical frameworks. Nature in these systems was sometimes lawful, sometimes unruly, and sometimes simply the everyday world resulting from nature's creativity. Many of the significant terms were gendered male or female and associated with ideas of order/disorder; form/matter; intellect/receptacle; being/becoming; lawful/unruly, and so on.

Early Greek Ideas of Nature

The foundational questions for Greek (and other) philosophers were:

1 Of what is the world composed and how does change occur? The Ontological Question.
2 How do we know? The Epistemological Question.
3 How should we act? The Ethical Question.

The ontological question involves nature as a creating principle (*natura creans/naturans*) and nature as created world (*natura creata/naturata*). The created world is composed of entities such as ideas or matter (or particles of matter); change occurs through processes such as those brought about by a creator, a god or gods, a demiurge, by emanation from the original One, or as the ceaseless activity of atoms or corpuscles. The epistemological question involves knowing, e.g., (for Plato) through recollecting what was lost to the individual mind after birth, by comparing the world of appearances to the pure forms, through ideas such as mathematics or the Good, or (for Aristotle) by observation through the senses. The ethical question involves the choices that humans make in leading the good life and how those choices can be justified.

24 *Autonomous Nature*

These questions set up major problems for Greeks, Romans, early Christians and beyond. Especially important were dichotomies and dialectics that existed between the concepts of order/disorder; stability/chaos; lawful/unruly; permanence/change; perpetuity/turmoil; creation/annihilation; good/evil. How did nature as an active, creative force come into conflict with nature as an unchanging, lawful order? What early ideas on the formation of the created world existed and how and when did the problem of nature as unruly and rebellious arise?

The ideas of nature and law emerged early in Western history. The terms *Phusis* (*Physis*), nature, and *Nomos*, law, presented a dichotomy containing embedded gendered meanings with women often being associated with *physis* (feminine) and men with *nomos* (masculine). Here the contradictions between nature as an active force and nature as a lawful order begin to arise.[4]

Two fundamental contradictions identified by early Greek thinkers laid the foundations for nature as both unchanging and changing, predictable and unpredictable, ordered and chaotic. The first idea of permanence and stability was expounded by Parmenides of Elea (fl. ca. 500 B.C.E.), the second idea—the lack of permanence and the continuance of change—was elaborated by Heraclitus of Ephesus (ca. 535–475 B.C.E.).[5]

Parmenides flourished in Elea on the coast of southern Italy around 500 B.C.E. Using a series of logical noncontradictory statements, he argued that Being exists and change cannot exist. By its very definition, Being is and Not-Being is not. Existence exists and nonexistence does not exist. Moreover, Being is eternal. It has always existed and has no beginning or end. Change does not exist and is, in fact, impossible. Why? Because Being cannot come from Not-Being, because Not-Being by definition does not exist. Logically, A (i.e., Being, a thing) cannot arise from not-A (i.e., Not-Being, no-thing). Nor can Being be destroyed, because by definition it cannot go into nonexistence to become Not-Being. Nor can Being arise from something, because if Being came from Being, how would it differ from its parent? It cannot vanish, for to do so it would become Not-Being, but Not-Being by definition does not exist.

Thus Being is, has always been, and always will be. There is no time; no past, present, or future. Being is one, continuous and without parts. For how could the parts of Being be distinguished from each other? One part could be distinguished from another only if some Not-Being existed beside it (as if cut into pieces by a biscuit cutter). But Not-Being does not exist (and therefore cannot act as a biscuit cutter). We define a thing by what it is not. But we cannot define Being by what it is not because Not-Being does not exist. Being is therefore without parts and continuous throughout. It is One. The unifying first principle of philosophy, therefore, is the One. The One would be interpreted by Plato as the unchanging pure Form, by Christianity as God, and by the mechanistic philosophy of the Scientific Revolution as God's unchanging laws of nature.

Around the same time, i.e., 500 B.C.E. Heraclitus of Ephesus deposited at the temple of Artemis his untitled book containing all his knowledge. Among its maxims are: "Nature loves to hide"—Nature hides "her" secrets from humans. And, "You cannot step twice into the same river for fresh waters are ever flowing in upon you,"—Everything is in constant flux and ever changing. From these phrases emerge many of the complex, contrasting meanings associated with nature as an unpredictable, ever-changing, chaotic force versus nature as a predictable, lawful order.[6]

Heraclitus's phrase "nature loves to hide" implies that nature's secrets are hidden from human knowledge and must be discovered. Artemis, at whose temple Heraclitus deposited his book of knowledge, like her Egyptian counterpart Isis, personified a Nature who harbored secrets to be revealed, unveiled, and extracted in the search for insight and understanding. Extracting those secrets, or truths, from an active, creative nature (*natura naturans*) would allow not only for knowledge of the natural world (*natura naturata*), but also for the possibility of its control and management. Mathematical analysis of pure forms and logic was one approach; experimentation and technology (*techne* or art) the other. Plato's divine Craftsman exemplifies the first approach, Hippocrates' earthly Craftsman the second.

Plato's Demiurge as Divine Craftsman

Plato's (424/3–348/7 B.C.E.) account of the creation of nature comes from the *Timaeus* (*On Physis*), one of his later dialogues. Here the tensions between creating and created, order and disorder, being and becoming, quantities and qualities play out in the form of idealism. Creating is done by an Intellect (*Nous*)—the Demiurge (*dêmiourgos*) or Divine Craftsman—who creates the universe (*kosmos*) or world by imitating eternally existing, unchanging forms. The result is the ever-changing world of appearances, composed of imperfect copies of the pure forms. The pure forms are patterns (*paradegmata*) from which comes the term *paradigm*. Socrates engages Timaeus in a discourse on "that which always is and has no becoming [i.e. being]" and "that which is always becoming and never is [i.e., change]."[7]

In the dialogue, Socrates questions Timaeus about the way in which the world came to be. Initially, the world existed in a state of disorder and chaos (*khaos*). The cosmos (or world) was created by the Demiurge, an intelligent craftsman, or Intellect (*Nous*)—the father, who, by imitating unchanging forms, brings order to the chaos. Prior to the world created by the Demiurge, everything is in disorder and disarray: "Finding the whole visible sphere not at rest but moving in an irregular and disorderly fashion, out of disorder, he brought order. . . ."[8]

The created world is an intelligent, living animal consisting of a female World Soul (*psyche*) and a receptacle (*chora*), or material substratum (*hyle*)—the mother. The receptacle is formless and female with only traces of the four elements out of which the Demiurge constructs the world.

> Wherefore, the mother and receptacle of all created and visible and in any way sensible things, is not to be termed earth, or air, or fire, or water, or any of their compounds or any of the elements from which these are derived, but is an invisible and formless being which receives all things. . . .

The created world that results is the world of appearances. It is notable with respect to subsequent intellectual and environmental history that the Intellect is male and the material receptacle is female.[9]

26 *Autonomous Nature*

The world the Demiurge creates is as much like himself as possible and based on an eternal, unchanging model from which order is imposed on the pre-cosmic chaos. That order comes into existence by Necessity (*ananke*) from the unchanging properties of its mathematical elements or perfect solids. Of the perfection and goodness of the Creator (*dêmiourgos*) in relation to the imperfectly created world, Plato writes:

> Now everything that becomes or is created must of necessity be created by some cause, for without a cause nothing can be created. The work of the creator, whenever he looks to the unchangeable and fashions the form and nature of his work after an unchangeable pattern, must necessarily be made fair and perfect; but when he looks to the created only, and uses a created pattern it is not fair or perfect. . . . [A]ll sensible things are apprehended by opinion and sense and are in a process of creation and created.[10]

The world's material body comprises the elements fire and earth, mediated by air and water, each of which has a perfect geometric shape [tetrahedron (fire), octahedron (air), icosahedron (water), and cube (earth), each formed from triangles]. The World Soul and material body are joined together as an organized whole comprising all living things. These things, however, are imperfect as they are formed from cast-off portions of the soul that get imbedded in matter. They are disturbed and confused by impacts among the parts. Each soul created by the Demiurge borrows pieces of earth, air, fire, and water and melds them together to form sensible bodies. The resulting created world is corporeal, visible, tangible, and in constant flux.

In some instances, however, Necessity resists the Intellect's (Reason's) persuasion resulting in a less perfect world that can be unstable. Here the element of uncertainty and potential rebelliousness, manifested in later Neoplatonic accounts, is intimated:

> Mind, the ruling power, persuaded necessity to bring the greater part of created things to perfection, and thus and after this manner in the beginning, when the influence of Reason got the better of Necessity, the universe was created.[11]

Plato thus answers the fundamental questions of creating and created, of order and change. His account depends on the theory that the eternal, unchanging mathematical forms and ideas are the patterns on the basis of which the Demiurge creates the existing world from a previously existing, chaotic disorder.

Thus in Greek cosmologies, both chaos (*khaos*) and the receptacle (*chora*) are female, while the Demiurge and the patterns or pure forms/ideas are male. For Plato, the change and uncertainty of Heraclitus are reflected in the created world, i.e., the sensible world of appearances; the certainty and permanence of Parmenides in the pure forms exemplified by mathematics. These ideas would evolve into the eternally existing, uncreated creator, or male God of Christianity and the female, sometimes unpredictable and vindictive Earth Mother of the Neoplatonists and Renaissance. In order to understand, predict, and hence manage the created world, mathematics is the key component. Plato thus anticipates the mathematical God of the Scientific Revolution and the 1687 *Principia mathematica* of Isaac Newton. His account of the Demiurge as craftsman, however, differs from the craftsman of the Hippocratic corpus. In the Hippocratic treatise, *On the Techne*, the craftsman manipulates matter in the created world to effect a cure, anticipating the seventeenth century's experimental method envisioned by Francis Bacon.

The Hippocratic Craftsman

The Hippocratic corpus, following from its founder, Hippocrates (460–370 B.C.E.) holds clues to the origins of the Greek view that nature's secrets could be extracted from the created world through art, i.e., technology (*techne*). Heinrich von Staden has discussed the meanings of *physis* and *techne* in the Hippocratic corpus. For Hippocrates and his followers, *physis* meant the discovery of regularities in the natural world and in the body. Rational design and behavior were recognizable characteristics of an individual or a disease. The *physis* or nature of a thing or a human being included not only its physical appearance, but also its powers and capacities. Practitioners should know both the nature of an individual and the powers of foods and drinks in order to reduce the strength of those that were too forceful or to increase them if they were too weak. "The physician's *techne* therefore constantly had to bear in mind not only a multiplicity of things with generalizable 'natures' and 'capacities' pertinent to every individual case of illness, but also the pervasive presence of particular swerves from the various norms expressed by nature (*physis*)."[12]

In some cases, however, the relationship between *physis* and *techne* was far more confrontational, leading to a process of "forcible constraint by human *techne*." In such cases nature does not yield the clues that could lead to a diagnosis and these must be forcibly extracted in order to perform the art of medicine. It is the *techne* itself that invents the tools of extraction. Von Staden has analyzed the Hippocratic treatise *On the Techne* ("On the Art" or more simply, "The Art"), written at the end of the fifth century B.C.E. Expressed therein is the view that "the *techne* at times should violate an unwilling, recalcitrant nature in order to force it to yield signs of the invisible. . . . " This is "a language of violence, of force, and of compulsion," not the view that *techne* is "a servant of nature, imitates nature," or "is a benign, cooperative extension of nature." Instead it reveals "the use of violent compulsion by *techne* against nature." The author of *On the Techne* writes:

> When this information is not afforded, and nature herself will yield nothing of her own accord, medicine has found means of compulsion, whereby nature is constrained, without being harmed, to give up her secrets; when these are given up she makes clear to those who know about the art, what course ought to be pursued.

28 *Autonomous Nature*

Here we can see the beginnings of an experimental approach that would ultimately constrain nature under controlled conditions for the purpose of understanding its secrets for the ultimate benefit of humankind.[13]

Although not advocated by Hippocrates or the author of *On the Techne*, in the early third century B.C.E., Herophilus and Erasistratus used dissection to wrest secretive information from human bodies. Under normal circumstances, the body's protective skin blocks its internal organs from view. Hence cutting a deceased body open was a method of prying out its secrets. More questionable was vivisection done on criminals to reveal the coloring and appearance of the living organs. Strong objections were mounted, based not only on human cruelty, but the distortion of the *physis* itself wrought by technological intervention. Vivisection not only violated nature but violated the *techne* itself. As von Staden observes,

> [N]ature both conceals and reveals, it veils and unveils, it closes and discloses. In particular, the invisibility of internal diseases and parts, which prompted the author of *On the Techne* to advocate forcing nature by artificial constraints to disclose its invisible features, was a frequent source of concern and complaint among medical writers. . . . [14]

Theodor Gomperz argues that *On the Techne* (*Peri technes*) presents a means of interrogating nature that anticipates the language of Francis Bacon. Nature is subjected to a painful interrogation in which she is forced to give up answers. He likens the process to the idea of torture, as in the German phrase *Folterzwang* or "coercion by torture," and to *Zwangsmittel* (coercive means). From a medical perspective, putting the body in traction is a means of compulsion that forces it to give up its secrets. Surgery is likewise akin to a legal procedure in which interrogation is used to extract evidence from an unwilling witness in the courtroom. A silent, recalcitrant witness must be forced to betray its own secrets. Gomperz compares the Greek treatise, *On the Techne*, to Bacon's *De Augmentis*, Scientiarum, 1623 quoting the phrase in which Bacon refers to Proteus under constraint.

> For like as a man's disposition is never well known or proved till he be crossed nor Proteus ever changed shapes till he was straitened and held fast, so nature exhibits herself more clearly under the trials and vexations of art (*natura arte irritata et vexata*) than when left to herself.[15]

In addition to the *De Augmentis*, Gomperz cites Bacon's 1620 *Novum Organum* which states that "so likewise the secrets of nature reveal themselves more readily under the vexations of art (*vexationes artium*) than when they go their own way." He also cites Bacon's "Plan of the Work" to the *Instauratio Magna* (1620), where he writes of "nature under constraint and vexed" (*constrictae et vexatae*) when by "art and the hand of man (*per artem et ministerium humanum*) she is forced out of her natural state and squeezed and moulded" (*premitur et fingitur*) and in which "the nature of things betrays itself more readily under the vexations of art (*vexationes artis*) than in its natural freedom." Finally, he quotes the 1609 *Wisdom of the Ancients* where Bacon writes: "Nevertheless if any skillful servant of nature

Here Gomperz is arguing that the Hippocratic author was ahead of his time, but did Bacon, in his use of the phrase, "vexations of art" (*vexationes artium*), know of the treatise, *On the Techne* (*On the Art*)? The phrase "vexations of art" has been translated and interpreted by many of Bacon's followers over the years as meaning the tortures or torments of art (technology). *On the Techne* was, in fact, available in the sixteenth century through the Latin translations of Marcus Fabius Calvus in 1525 and Janus Cornarius in 1538 and 1546, so Bacon may well have known of the treatise and drawn ideas and language from it.[17]

Here technology (*techne*) serves as a method for manipulating the created world through human intervention and hence to managing and controlling it through extracting its secrets. This approach would lead to the experimental method of the seventeenth century scientists.

Aristotle on Creating and Created

Aristotle's (384–322 B.C.E.) *Physics* (*Physis*) builds on, but differentiates itself from preceding accounts by asserting that the real world is the empirical world of sense perception. The Form does not exist in a separate world of pure Forms, as for Plato; rather form and matter exist together within each object. *Physis* and *Techne* (Nature and Art) combine differently to produce the objects of the real world. Each individual thing reflects a combination of four causes: material, formal, efficient, and final. The way in which the creating process produces the created object, however, differs in Nature (*Physis*) and in Art (*Techne*).[18]

In objects of living nature, the creating process by which living things are generated and grow lies in the way the four causes operate together. The material cause of a particular object is its own matter (egg, seed, embryo). The formal cause comprises an object's form or pattern (including its shape, size, color, and smell). But the form exists only in conjunction with matter, and matter exists only in conjunction with form (they can be separated in thought, but not in actuality). Matter and form in living things are gendered: matter for Aristotle is female (the egg), while form is male (the sperm).

In living objects, the formal, efficient, and final causes interact such that a material object actualizes the potential within it to reach its final form. Thus an acorn becomes an oak tree, an egg becomes a chicken, a baby becomes a human. The formal cause is the form or pattern (shape, size, color, smell, taste, texture) attained by the mature living thing (the tree, chicken, or human). The efficient cause is the principle of growth within the seed or embryo. The final cause is the goal it strives to attain—i.e., the mature plant or animal (the actualization of the potential in its seed or embryo). In nature, therefore, the creating process that exists within nature "herself" produces the created object. Here, Nature is the creating artist.

In art, objects (such as a bed, table, or urn), are created by an artist, craftsman, or technician. The matter is the wood or clay, which is some combination of the four elements (earth, air, fire, and water). The formal cause is the shape and size of

30   *Autonomous Nature*

the bed, table, or urn and its associated qualities (firmness, softness, color, texture, and decoration). The efficient cause is the artist or technician who takes the matter and creates the object—the bed or table. The final cause or goal is the pattern or purpose existing in the mind of the artist—the purpose of a bed is a place to sleep, that of an urn to hold liquid. In things of art, the four causes operate separately, but in combination produce the object itself. In art, the creating process thus exists in the person of the artist who combines the matter, form, and purpose to produce the created object.

Aristotle's philosophy of nature explains the existence of the cosmos and its *telos*, or purpose. The ultimate cause of the cosmos is the unmoved mover, uncreated and existing eternally. There is no initial cause or first motion, only a final cause; no *creatio ex nihilo* (as there would be in Christianity), but eternal existence. All things strive toward the unmoved mover as the ultimate final cause—the end and purpose of their existence. There is no void space; the universe is a plenum with the earth at its center. Below the moon everything is composed of some combination of the four elements, earth, air, fire, and water. Above the moon, the rest of the cosmos consists of the fifth element, ether, out of which the stars and planets are formed. Each of the four elements has a natural place in a hierarchy with earth at the center, then water, then air, and finally fire.

Unpredictability and disorder in nature arise in Aristotle's philosophy via chance and spontaneity in opposition to necessity. In nature, almost everything happens by necessity. Related to self-active nature, however, are the functions of chance and spontaneity. How, asks Aristotle, can errors in nature be explained? Although all things in nature exist for a purpose and strive toward completion of that purpose, mistakes occur. Monstrosities result from a failure of the original purpose. Things of nature which fail "to reach a determinative end must have arisen through the corruption of some principle corresponding to what is now the seed." "In natural products the sequence is invariable, if there is no impediment." The world is thus a rational order, operating according the laws of necessity, but one into which chance and spontaneity can inject irrationalities and disorder.[19]

Aristotle's cosmos, therefore, is made up of the five elements—earth, air, fire, water, and ether. Change occurs by the striving of each object to reach its final form—ultimately the first cause or unmoved mover. The unmoved mover would later be synthesized with the God of Christianity, as first cause, to become the worldview of the late medieval period manifested in the philosophy of Thomas Aquinas. It would be challenged during the Scientific Revolution of the seventeenth century which retained the material (matter), formal (quantities, shape, size), and efficient (force) causes, while eliminating qualities (color, sound, flavor) and final (teleological) causes. In the seventeenth century, philosophers such as Pierre Gassendi (1592–1655) and Isaac Newton (1642–1727) would revive the atomistic philosophies of Democritus, Epicurus, and Lucretius (see below) along with the idea of void space, while others such as René Descartes (1596–1650) and Thomas Hobbes (1588–1679) would argue that material corpuscles existed in a plenum in which no vacuum (void space) could exist.

Yet Aristotle also presages another significant development relevant to the distinction between *Natura naturans* and *Natura naturata*—that between nature as art and the human as artist. For Aristotle, art (*techne*) was the imitation (*imitatio*) of nature. He wrote, "art partly completes what nature cannot bring to a finish; and partly imitates her." The carpenter can complete what nature would do if "she" had the capability. What nature began by growing trees, the human builder of beds and houses can finish for "her," because nature "herself" cannot do so. What is potential in wood can therefore be completed by the artist/craftsman. The craftsman who has the end (final cause) in mind creates a design (formal cause) and uses human skill (efficient cause) to unite the form with the matter to achieve the human end and thus complete the series. Here *techne* means the production of artificial objects. Nature is both a productive principle of growth (*natura naturans*) and a produced form (*natura naturata*).[20]

While the achievements of Plato, Aristotle, and the Hippocratic authors set the stage for the predictability and control of nature through mathematics and experimentation, those of the Roman philosophers, Lucretius and Plotinus, instead anticipate nature's unpredictability, unruliness, and unlawfulness. Here the concepts of chaos and complexity fundamental to the sciences of the twentieth century make their debut. Incorporating earlier ideas of chaos, chance, and spontaneity, Nature now manifests a side of "herself" that is free, willful, rebellious, and unlawful. What do these two philosophers argue and why are they important for "autonomous nature"?[21]

Lucretius on *Natura Creatrix*

The Roman poet, Titus Lucretius Carus (99–ca. 55 B.C.E) is best known for his six-book poem, *De Rerum Natura* ("On the Nature of Things"). His dates are uncertain, but Cicero knew of his poem by 55 C.E., commending it for its "many flashes of genius" and "great mastership." Lucretius was a follower of the Greek philosopher, Epicurus (third century B.C.E.), whose book was titled *On Nature* (*Peri Phuseos*), and he built on Epicurus's account of the created world as made of atoms and void space. Lucretius, however, titled his book *De Rerum Natura* (rather than simply *De Natura*) to explain both atomic reality and the phenomenal world. The book was recovered in Renaissance Florence in the early fifteenth century and was influential thereafter. Change occurred, not only by random motions, but also by the actions of the *creatrix* (creatress), *Natura*. Lucretius dealt with fundamental topics such as Being and Not-Being, matter and void, infinity and finitude, time and space, mind and spirit, mortality and immortality. Books 1 and 2 cover matter, space, and the motions of atoms; Books 3 and 4 the nature of human beings; and Books 5 and 6 the nature of the cosmos itself.[22]

Lucretius described the dichotomy between the creative activity of nature and its willful uncontrollability in his *De Rerum Natura*. Two features stand out with respect to the history of predictability and control. On the one hand, Nature is personified as a creatress (*creatrix*), who as active force creates the phenomenal world

32   Autonomous Nature

from atoms in ceaseless motion. "At length everything is brought to its utmost limit of growth by nature, the creatress (*natura creatrix*) and perfectress." On the other hand, as *natura libra*, she is free, uncontrollable, and capable of action without the supervision of any gods (later God): " . . . you will immediately perceive that nature is free and uncontrolled by proud masters and runs the universe by herself without the aid of Gods."[23]

It is the tension between order and disorder that underlies the problem of nature as law, order, and stability (hence rational and predictable), and nature as disorder, willful, and unruly (therefore unpredictable and uncontrollable). Nature is both creator and dissolver, generator and disrupter, rational and irrational, consistent and inconsistent.

Robert David Clark has written extensively about *Natura creatrix*, or nature as agent, active force, and actress in Lucretius's *De Rerum Natura*. More than any other Roman writer, Lucretius established *natura* and nature's personification as a female creative force (*Natura*) central to the Latin world and to an understanding of nature itself. Lucretius posits atoms in random motion without direction or purpose (*rerum natura*) falling through space, deflected by an inherent swerve, but, it is *Natura* who organizes the atoms to form the created world. In earlier Greek accounts, Venus was the generator or *genetrix* of nature, giving birth and nourishment, reproduction and regeneration, to a fertile earth, but the Roman *Natura* becomes its universal *creatrix*. *Natura*, moreover, takes on the functions of governor (*natura gubernans*) or supervisor who steers the winds and tames the rolling seas. The regenerative aspect of nature is part of *Natura*'s persona as well as her freedom (*libra*) to act without control.[24]

Clark writes: "Since the Epicurean world is one which is produced by chance . . . [Lucretius's] description of 'governing' forces . . . as well as the use of natural imagery to stress the regularity, predictability, and *rationality* of the universe is noteworthy."[25]

The creative and destructive aspects of nature are not dichotomies, but a dialectic—two sides of *Natura*'s coin of activity. *Natura* takes the uncreated, eternally existing atoms and as agent fashions them into the natural world. Similarly, when the things of nature are destroyed they are dissolved (rather than destroyed) by *Natura* back into eternally existing atoms. Creation and dissolution operate in tandem. *Natura* is a creative unity, overseeing a rationally understood, lawful and orderly process of creation and dissolution. Atoms exist eternally; they are not created *ex nihilo*. If created out of nothing, then anything would be possible. Humans could be born from the earth, wild animals could inhabit farmlands, and trees could bear all kinds of fruit.[26]

Natura fuses the activities of the aethereal heavens (*aether*) and the fruitful earth (*terra*). Rain from the heavens (*pater aether*) nourishes and gives rise to new life on earth (*mater terra*). The sky father's celestial semen fertilizes the earth mother. In the process of creation, however, nature not only encourages and makes possible the birth of new things, "she" also intervenes, using force on recalcitrant matter to make it produce. She forces atoms to be bounded by void and void to bound each individual atom. It is *Natura*'s intervention that allows for certainty and predictability in the continuance of life. She designs and controls the universe. She embodies law and limits, creating an orderly, rational, predictable natural world.[27]

In Books 5 and 6, however, a wilder and more uncontrollable persona of *Natura* emerges. Here Lucretius describes the earth as made up of wild animals, swamps, and thorns. Boisterous seas, earthquakes, and tornados characterize nature's activity. But *Natura* also plays a moderating role. When Earth (*terra*) tries to bring forth monsters with distorted appearances and missing body parts (as would be the case in an Epicurean world of atoms in ceaseless, random motion), she is checked by *Natura*'s more rational power. That rational power is expressed as the laws of nature (*foedera naturae*). The laws of nature constrain the appearance of monsters and limit the unpredictability of natural phenomena. The violence of anarchy and randomness of chance that might appear in an Epicurean universe are held in check in Lucretius's universe by law.[28]

Lucretius uses the term *foedera naturae(i)* (laws of nature) to characterize the lawfulness and regularity of nature. In Book 1, he states: "Finally, a certain limit has been ordained for the growth and life-span of specific things; and the unyielding laws of Nature have decreed what each thing can do, and what it cannot do." In Book 2 he writes: "And whatsoever things have been born will be born again in the same manner, and they will grow up and flourish as far as each is permitted by the laws of Nature." It is because of the laws of nature that species maintain their own genres and do not mingle to produce hybrid monsters, but "each grows in its own way, and all of them retain their properties according to Nature's laws." In Book 5, he notes that not even the crumbling rocks or shrines of the gods can "combat the laws of nature."[29]

In Book 2 (repeated in Books 3 and 6) he uses the phrase *naturae species ratioque* (aspect and law of nature) to express the rationality of nature: "This terror of the mind, these shadows must be dispelled not by the sun's bright shafts nor by the brilliant daylight, but by an understanding of the laws of Nature." And in Book 5, he uses that phrase to identify himself as the first Roman to explain the laws of nature: "the very laws of nature have only now been discovered—in fact, I am the first who has been found to explain these truths in the Latin tongue."[30]

In the *De Rerum Natura*, therefore, the two aspects of *Natura* as an active, creative whole come together—the first of nature as the creatress (*creatrix*) and perfectress of a lawful, rationally created order, the second of nature as the free, uncontrolled dissolver of that same creation. It is these two aspects of nature as active creating force—both orderly and disorderly, lawful and willful—and the resultant nature as created world that would be taken up by Plotinus and the early medieval Christians.

Plotinus on Creating versus Rebellious Nature

Plotinus (204–270 C.E.), a follower of Plato and founder of Neoplatonism (so named in the nineteenth century), was a major contributor to an idealist philosophy of nature. After studying in Alexandria for eleven years and traveling to Persia, he settled in Rome in 245 C.E. where he lectured and wrote the essays and

34 Autonomous Nature

notes that would later be complied by his student Porphyry as the *Enneads*. His philosophy of the reality of ideas and forms as opposed to the transitory world of the senses became an important influence on Christian thinking. St. Augustine in particular found a deep resonance between Plotinus's philosophy and Christian ideas. Plotinus's foundational concepts are the One (*Monad, monadikos* [male]), the Intellect (Male, *Nous*), and the Soul (Female, later the Latin, *Anima*) which had two portions, the lower portion being Nature (*Phusis*, later the Latin, *Natura*). The world was created not from nothingness (*ex nihilo*) but by a process called emanation.

Plotinus's philosophy is critical to the problem of change and how change occurs. Parmenides's concept of Being as One, indivisible, unchanging, and permanent did nothing to explain the sensible world. For those who agreed, however, that the One was a transcendent unchanging unity, the issue of the emergence of the created world became a central issue. For Plotinus that emergence was explained in two ways. The first was the process of emanation, or overflowing, the second the process of *tolma*, or willfulness. Emanation led to an explanation of the created world as one of order, beauty, and good, deriving from the unchanging ideal form of the One. *Tolma* explained change as disorder, chaos, and turmoil. Both processes were necessary to explain the activity of the world of nature.

As a follower of Plato, Plotinus held to the philosophy that reality was manifested in unchanging pure forms or ideas. Ideas did not change, but the sensible world of appearances was an imperfect reflection of the pure form. Mathematics expressed the pure form most clearly. While the pure form of the triangle held in the mind was of three lines and three angles, all triangles in the created, sensible world were imperfect owing to their matter. Even the purest, clearest, most perfect triangle in the world of appearances was still made of matter and hence impure and imperfect. Order, Beauty, and the Good as pure forms were only imperfectly reflected in the world of human life and nature. For Plotinus, the problem of maintaining the integrity of reality as unchanging pure form, or the One, conflicted with the problem of explaining the creation of the ever-changing imperfect world of nature.

The problem of creating (later identified as *natura creans*) as a process and the problem of the created (*natura creata*) that resulted from the creating process were at the root of Plotinus's inquiry into nature and the way that the natural world emerged from the unchanging One. It is here that the principles of emanation to explain order and lawfulness and *tolma* to explain disorder came into play.

Plotinus's concept of emanation is most clearly explicated in *Ennead* 3, viii, titled, "On Nature, Contemplation, and the One." The fundamental concepts of his philosophy are the One, the *Nous* (Intellect), and the Soul. The One is light—transcendent, indivisible, immutable, and beyond all Being and Not-Being. It is synonymous with Good and Beauty. The cosmos unfolds out of the One, out of the light that keeps on shining eternally, without creation, and without diminution. From it emanate light rays which are always less perfect than the original source itself. The first emanation from the One is Nous—intellect, mind, reason, thought, order, logos. Nous is akin to the light of the sun. The next emanation, the

Greco-Roman Concepts of Nature   35

World Soul, is like the moon whose light derives from the sun. The World Soul has an upper and lower portion, the lower part of which is Nature. From Nature contemplating "herself" emanates the natural world in stages of decreasing perfection. The created world, however, is still divine because it emerges from the original divine One.[31]

The intellectual World Soul contemplates a sublime spectacle. Of the Soul's process of contemplation, Plotinus writes:

> Since then we have considered in what manner the fabrication of nature is a certain contemplation, let us next proceed to that soul which is superior to nature; for the contemplation of this soul, its ingenuity, its desire of learning and inquiry ... produces a parturient and abundant fecundity, so that becoming a spectacle throughout it generates another spectacle. . . . [32]

The created world is alive. Energy runs through all things, producing life from life itself. Yet what is produced is gradually weakened by its descent from the original contemplation.[33]

The created world thus arises from Nature's contemplation of herself. Nature (the lower part of the World Soul) operates through her own energy; from her emanate the outlines of bodies in the same manner as a geometrician contemplates figures. But rather than engraving bodies in matter, she lets them emerge from her own thought process. Nature is not like a craftsman who produces images from wax by pressing and molding, but instead fabricates them without the use of appendages. She herself is pure form and not a composite of form and matter. She operates purely by speculation through reason, producing spectacles, akin to mathematical theorems, ready to behold. Matter, as the subject of the fabrication, is acted on by form.[34]

Plotinus used metaphors from nature to explain the process of emanation. Thus the energy of life emanates from the One as rivulets from an inexhaustible fountain that already knows the directions of the derivative streams.

> Conceive then a fountain possessing no other principle, but imparting itself to all rivers, without being exhausted by any one of them, and abiding quietly in itself; but the streams which emanate from this fountain, before they flow in different directions, as yet abiding together, and as it were, already knowing what rivulets will proceed from their defluxions. . . . [35]

Another metaphor from nature was taken from a tree whose roots sustained its life, but was not itself divided or diminished in the process of multiplication.

> [O]r conceive the life of a mighty tree, propagating itself through the whole tree, the principle at the same time remaining without being divided through the whole, but as it were, established in the root: this then will afford an universal and abundant life to the tree, but will abide itself, without multiplication and subsisting as the principle of multitude. . . . [36]

36   *Autonomous Nature*

While emanation explained the process of change that maintained the integrity of the One and its order, beauty, and goodness, another process was needed to explain the manifest disorder in the created world. That process was *tolma*, or willfulness.

In explaining the creation of the natural world as imperfect and disorderly, Plotinus described the fall of the Soul from the original, supreme One. The fall is triggered by the Soul's own desire to desert its First Principle and to assert itself, the process of *tolma*. This word has been translated as the Soul's willfulness, boldness, and "rebellious audacity." The Soul has its own motive that leads it to forget and desert its maker. Its fall stems from a desire to have a life of its own. Individual souls continue the process of degradation, forgetting their origins in their wish to unite with matter.[37] Plotinus writes:

> The evil that has befallen them is due to a Rebellious Audacity . . . They begin to revel in free-will . . . Smitten with longing for the Lower, rapt in love for it, they grew to depend upon it: so they broke away, as far as was in their power and came to slight the lofty sphere they had abandoned.[38]

Plotinus describes the *tolma* elsewhere, stating that "The first Dyad separated itself from the Monad in what is called an act of rebellion or self-assertion, a *tolma*." He also writes that "in plants there dwells, 'the more rebellious and self-assertive part (or phase) of the soul.'" In fact, even the Nous or Divine Mind in the first emanation from the One does so "in an act of self-assertion." Plotinus makes a further distinction between freedom and willfulness. Freedom is right action; willfulness is an act driven by the senses, by a law of nature.[39]

The particular souls are all part of the unity of the one Soul, forming a multiplicity within the greater unity which is itself part of the Intellect (*Nous*). The Soul is a creative principle. In its creative exuberance, it overflows, ensouls, and breathes life into

> all that the earth nourishes and the sea; all that are in the air and all the divine stars in the heavens; itself has formed the sun and this vast firmament of sky . . . and it is a Nature apart from all which it gives the order and the movement and the life . . . as the rays from the sun pour light upon a gloomy cloud and make it shine in a golden glory. . . .[40]

As a result of the fall of the souls, however, a return is needed. That return can only be accomplished by stages of purification. The souls must first recognize that the new world they have embraced (nature) is shameful. They must be reminded that it is the One that is to be honored. They must remember that each living thing was a dead body before the soul entered it.[41]

Plotinus's sensible world, the world of nature, is a living thing, reflecting both order and disorder, good and evil, perfection and imperfection. Everything is alive, including the sun, stars, and heavens. Matter as foundation and substratum takes on individual shapes, giving particularity to the world. The sensible world contains a vastness, beauty, and harmony derived from the One. From *tolma*, however, derives the change and disorder manifested in the existing world.[42]

In characterizing Plotinus's relevance to current concepts of chaos and order, ecologist Daniel Botkin writes that to change the ways in which we manage our resources,

> [W]e must confront the very assumptions that have dominated perceptions of nature for a very long time. This will allows us to find the true idea of a harmony of nature, which as Plotinus wrote so long ago, is by its very essence discordant, created from the simultaneous movements of many tones, the combination of many processes flowing at the same time along various scales, leading not to a simple melody but to a symphony at some times harsh and at some times pleasing.[43]

Change and permanence are both part of Plotinus's explanation of nature creating and nature created. Permanence, form, and oneness explain reality—the idealist philosophy of both Platonists and Neoplatonists. Change is needed to explain this world and how it was created out of the unchanging One. Emanation and *tolma* are the processes of emergence and change that account for the world around us. Willfulness, disorder, and discordance are integral parts of that change.

Conclusion

By the first two centuries of the Common Era, a philosophy had emerged from the Greco-Roman period that would go forward to shape the Christian world of the Middle Ages. How the idea of the One related to the Christian idea of God and how the ideas of *tolma* or willfulness in nature related to the Christian idea of moral evil would set the stage for the philosophies of Augustine, Johannes Scotus Erigena, Thomas Aquinas, and Bonaventure. From Greco-Roman philosophies, which drew heavily on Aristotle's *phusis* and Plotinus's *natura*, the Latin concepts of *natura creans/naturans* and *natura creata/naturata* (and their meanings and tensions) would develop. Especially important were the gendered associations of intellect, form, idea, and order with the male and chaos, matter, receptacle, and disorder with the female. These ideas would in turn influence the development of experimentation and mathematics as ways to predict and control nature during the Renaissance and Scientific Revolution.

Notes

1　Pliny the Younger, *The Letters of the Younger Pliny*, trans. with an introduction by Betty Radice (Baltimore: Penguin Books, 1969); BR.6, Letter 16, pp. 166–168, quotation on p. 168, Letter 20, pp. 170–173.

2　Hesiod, *Theogony*, trans. H.G. Eveyln-White, II, 116–122, http://www.greekmythology.com/Books/Hesiod-Theogony/Theog__116–206_/theog__116–206_.html. On Khaos as female progenitor, see also http://www.theoi.com/Protogenos/Khaos.html.

3　Hesiod, *Theogony*, ibid., ll. 176–177.

38 Autonomous Nature

4 Gerard Naddaf, *The Greek Concept of Nature* (Albany, NY: State University of New York Press, 2005), pp. 133, 161, 164. "The Greeks would refer to law, order, and rationalism as *nomos*. Women in general tend to be (in the view of Greek man) on the side of *physis*, while men are generally on the side of *nomos*. . . . Men make and obey laws, but women do what comes naturally (see Phaedra or Medea)," quotation from https://en.wikipedia.org/wiki/Physis.

5 Although contemporaries, Heraclitus apparently wrote somewhat before Parmenides as he does not mention the latter. On the other hand, Parmenides seems to refer to Heraclitus in his poem when he writes: "Helplessness guides the wandering thought in their breasts; they are carried along deaf and blind alike, dazed, beasts without judgment, convinced that to be and not to be are the same and not the same, and that the road of all things is a backward-turning one." See: http://www.iep.utm.edu/heraclit/ and http://en.wikipedia.org/wiki/Parmenides.

6 Pierre Hadot, *The Veil of Isis: An Essay on the History of the Idea of Nature*, trans. Michael Chase (Cambridge, MA: Harvard University Press, 2006), pp. 1–3.

7 Plato, "Timaeus," in *The Dialogues of Plato*, trans. Benjamin Jowett (New York: Random House, 1937 [1892]), 27d–27e.

8 Plato, "Timaeus," trans. Jowett, 30a.

9 Plato, "Timaeus," trans. Jowett, 27c–34a.

10 Plato, "Timaeus," trans. Jowett, 28a. See also 29a:

> [T]the world is the fairest of creations and he is the best of causes. And having been created in this way, the world has been framed in the likeness of that which is apprehended by reason and mind and is unchangeable, and must therefore of necessity, if this be admitted, be a copy of something. Now it is all-important that the beginning of everything should be according to nature.

11 Plato, "Timaeus," trans. Jowett, 48a.

12 Heinrich von Staden, "*Physis* and *Techne* in Greek Medicine," in Bernadette Bensaude-Vincent and William R. Newman, eds., *The Artificial and the Natural: An Evolving Polarity* (Cambridge, MA: MIT Press, 2007), pp. 21–49, quotation on p. 26.

13 Von Staden, "*Physis* and *Techne* in Greek Medicine," quotations on pp. 24, 30; Hippocrates, *The Art*, in *Hippocrates*, ed. and trans. W.H.S. Jones (London: William Heinemann, Loeb Classical Library, 1959), vol. 2, quotation on p. 215.

14 Von Staden, "*Physis* and *Techne* in Greek Medicine," pp. 32–37, quotation on p. 32; idem, "The Discovery of the Body: Human Dissection and Its Cultural Contexts in Ancient Greece," *Yale Journal of Biology and Medicine* 65 (1992): 223–241.

15 Von Staden, "*Physis* and *Techne* in Greek Medicine," pp. 30–31; Theodor Gomperz, *Die Apologie der Heilkunst: Hippocratis de Arte* (Leipzig: Veit & Comp., 1910), pp. 140–141. Gomperz's comparisons pertain to chapters 13–14 of *On the Techne*. I thank Heinrich von Staden for these citations and for assistance with the interpretation. Francis Bacon, *De Augmentis*, in *Works*, ed. James B. Spedding, Robert Leslie Ellis, and Douglas Devon Heath, 14 vols. (London: Longmans Green, 1868–1901), Bk. 2, Ch. 2, vol. 4, p. 298. Latin, cited by Gomperz, vol. 1, p. 500.

16 Bacon, *Novum Organum* (1620) in *Works*, op. cit, vol. 4, Bk. 1, Aphorism 98, p. 95 (Latin, vol. I, Bk. I, Aphorism 98, p. 203); Bacon, "Plan of the Work" (1620), *Instauratio Magna*, *Works*, vol. 4, p. 29 (Latin, vol. I, p. 141); Bacon, *Wisdom of the Ancients* (1609), *Works*, vol. 6, pp. 775–776 (Latin, vol. VI, pp. 651–652).

17 Marcus Fabius Calvus van Ravenna was an author and translator of the works of Hippocrates. He first translated the Hippocratic Corpus into Latin in Rome, 1525. See Ira M. Rutkow, *Surgery: An Illustrated History* (London and Southampton: Elsevier Science Health Science, 1993), p. 23. Janus Cornarius (1500–1558), *Hippocratis Coi medici vetustissimi . . . libri omnes* (*Complete Works of Hippocrates of Cos, Most Ancient of Physicians*), Latin translation (Basel, 1538); *Hippocratis Coi . . . Opera quae ad nos extant*

Greco-Roman Concepts of Nature 39

omnia (*The Extant Works of Hippocrates of Cos*), Latin translation (Basel, 1546). For the history of the interpretation of the phrase "vexations of art" as the tortures or torments of art, see Carolyn Merchant, "Francis Bacon and the 'Vexations of Art': Experimentation as Intervention," *British Journal for the History of Science* 46, no. 4 (2013): 551–599.

18 Aristotle, *Physics*, in *The Basic Works of Aristotle*, ed. Richard McKeon (New York: Random House, 1941), Bk. II, pp. 236–252.

19 Aristotle, *Physics*, ibid., 199a35; quotations on 199b6–7, 199b25–26. See also, "[T]he same completion is not reached from every principle . . . always the tendency in each is towards the same end, if there is no impediment" (199b17–19).

20 Aristotle, *Physics*, ibid., 199a15–17; Hans Blumenberg, "Toward a Prehistory of the Idea of the Creative Being," trans. Anna Wertz, *Que Parle*, special issue on *The End of Nature* 12, no. 1 (Spring/Summer 2000): 17–54, see pp. 17–18, 31.

21 Blumenberg, "Toward a Prehistory of the Idea of the Creative Being," pp. 17–18; Ilya Prigogine and Isabelle Stengers, *Order out of Chaos: Man's New Dialogue with Nature*, intro. by Alvin Toffler (New York: Bantam, 1984), p. 141; Richard Kautz, *Chaos: The Science of Predictable Random Motion* (New York: Oxford University Press, 2011).

22 See Lucretius, *De Rerum Natura* (1st c. B.C.E.), trans. W.E. Leonard (New York: E.P. Dutton, 1950); Lucretius, *The Nature of the Universe* (*De Rerum Natura*, 1st c. B.C.E.), trans. R.E. Latham (Baltimore, MD: Penguin, 1951); Lucretius, *On the Nature of the Universe (De Rerum Natura)*, a New Verse Translation [with line numbers] and an Introduction by James H. Mantinband (New York: Frederick Ungar, 1965); for the Latin text of Book 2 with a commentary, see Don Fowler, *Lucretius on Atomic Motion: A Commentary on* De Rerum Natura, *Book Two, lines 1–332* (Oxford, UK: Oxford University Press, 2002), pp. 242–244. For a discussion of the title of Lucretius's book as *De Rerum Natura*, rather than *De Natura*, see Robert David Clark, "*Natura creatrix*: The Matter of Meaning in the *De Rerum Natura*," PhD dissertation, Columbia University, 2000, p. 288.

23 Lucretius, *Nature of the Universe*, trans. Latham, Bk. 2, lines 1116–1117, p. 93; Bk. 2, lines 1090–1092, p. 92. See also Lucretius, *On the Nature of the Universe*, trans. Mantinband, Bk. 2, lines 1115–1116, p. 66, lines 1090–1092, p. 65.

24 Lucretius, *On the Nature of the Universe*, trans. Mantinband, Bk. 1, lines 1–43, 54–61; Bk. 2, lines 216–293. Clark, "*Natura creatrix*," op. cit., pp. 13, 23, 38–39, 40–42, 52.

25 Clark, "*Natura creatrix*," quotation on p. 60, Clark's italics.

26 Lucretius, *On the Nature of the Universe*, trans. Mantinband, Bk. 1, lines 149–158, 215–224, 174–192; Clark, "*Natura creatrix*," pp. 46, 53–54, 56, 58, 69.

27 Lucretius, *On the Nature of the Universe*, trans. Mantinband, Bk. 1, lines 250–252, 262–264; Clark, "*Natura creatrix*," pp. 76–78, 80–81, 95–96, 149–150, 153.

28 Lucretius, *On the Nature of the Universe*, trans. Mantinband, Bk. 5, lines 39–42, 105–106, 200–205, 335–337, 368–369, 837–848, 916–924, Bk. 1, lines 584–591, Bk. 6, lines 39–41; Clark, "*Natura creatrix*," pp. 235–237, 244–245, 353–359, 362; Charles William Gladhill, "*Foedera*: A Study in Roman Poetics and Society," PhD dissertation, Stanford University, 2008, see Ch. 2, "*Foedora* and Roman Cosmology," pp. 133–200.

29 Lucretius, *On the Nature of the Universe*, trans. Mantinband, Bk. 1, lines 584–587, p. 17 (Latin: "Denique iam quoniam generatim reddita finis crescendi rebus constat vitamque tenendi, et quid quaeque queant per *foedera naturai*, quid porro nequeant, sancitum quando quidem extat, . . . "). See also Lucretius, trans. Mantinband, Bk. 2, lines 300–302, p. 42 (Latin: "et quae consuerint gigni gignentur eadem condicione et erunt et crescent vique valebunt, quantum cuique datum est per *foedera naturai*"). In Book 5, the phrase *foedere naturae* appears twice: see Lucretius, *On the Nature of the Universe*, trans. Mantinband, Bk. 5, lines 306–309 (Latin: "Denique non lapides quoque vinci cernis ab aevo, non altas turris ruere et putrescere saxa, non delubra deum simulacraque fessa fatisci nec sanctum numen fati protellere finis posse neque adversus *naturae foedera* niti?") and 923–924 (Latin: "non tamen inter se possunt complexa creari, sed res quaeque suo ritu procedit et omnes *foedere naturae* certo discrimina servant").

40 *Autonomous Nature*

30 For the phrase "naturae species ratioque," see Lucretius, *On the Nature of the Universe*, trans. Mantinband, Bk. 2, lines 59–62, p. 35; Latin: "[H]unc igitur terrorem animi tenebrasque necessest non radii solis neque lucida tela diei discutiant, sed *naturae species ratioque*"; see also Bk. 3, lines 91–93, p. 70 and Bk. 6, lines 39–41, p. 179 where the phrase is repeated. See also Lucretius, trans. Mantinband, Bk. 5, lines 335–337, p. 145 (Latin: "denique natura haec rerum ratioque repertast nuper, et hanc primus cum primis ipse repertus nunc ego sum in patrias qui possim vertere voces").

31 Plotinus, "On Nature, Contemplation, and the One," *Enneads*, in *Collected Writings of Plotinus*, trans. Thomas Taylor (Frome, Somerset, UK: Prometheus Trust, 1994), *Ennead*, III. 8, pp. 123–134. For another translation see Plotinus, "Nature, Contemplation, and the One," in *Plotinus: Psychic and Physical Treatises; Comprising the Second and Third Enneads*, trans. Stephen Mackenna (London: Medici Society, 1917), vol. 2, *Ennead*, III. 8, pp. 119–136. For background, see Plotinus: *The Ethical Treatises*, trans. Stephen Mackenna (London: Medici Society, 1917), vol. 1, "Porphyry's Life of Plotinus," pp. 1–28, and "The Preller-Ritter Extracts Forming a Conspectus of the Plotinian System," ibid., pp. 130–158. Mackenna's translation of the "Extracts" is from H[einrich] Ritter and L[udwig] Preller, *Historia Philosophiae Graecae et Romanae ex Fontium Locis contexta* (Gothae, 1864). For commentaries, see John N. Deck, *Nature, Contemplation, and the One: A Study in the Philosophy of Plotinus* (Toronto: University of Toronto Press, 1967), esp. Ch. 6, "Nature," pp. 64–72; William Ralph Inge, *The Philosophy of Plotinus* (London: Longmans, Green, 1918), 2 vols.; N[atale] Joseph Torchia, *Plotinus, Tolma, and the Descent of Being* (New York: Peter Lang, 1993).

32 Plotinus, "On Nature, Contemplation, and the One," trans. Thomas Taylor, op. cit., quotation on p. 126.

33 Plotinus, ibid., pp. 126, 127.

34 Plotinus, ibid., pp. 124, 125.

35 Plotinus, ibid., quotation on p. 132.

36 Plotinus, ibid., quotation on p. 132.

37 Plotinus, *Enneads*, "The Preller-Ritter Extracts Forming a Conspectus of the Plotinian System," trans. Stephen Mackenna, vol. 1, *The Ethical Treatises*, op. cit., p. 130.

38 Plotinus, "Preller-Ritter Extracts," trans. Mackenna, ibid., p. 130 (note: selection is from *Ennead*, V. 1 but is incorrectly cited on p. 130 as V. 3.9). See Plotinus, *Ennead*, V. 1. For the Thomas Taylor translation of *tolma* as *audacity*, see *Collected Writings of Plotinus*, op. cit., "On the Three Hypostases that Rank as the Principles of Things," *Ennead*, V. 1, p. 299:

> What is the reason that souls become oblivious of divinity, being ignorant both of themselves and him, though their allotment is from thence, and they in short partake of God? The principle therefore of evil to them is *audacity*, generation, the first difference, and the wish to exercise unrestrained freedom of the will. When, therefore, they began to be delighted with this unbounded liberty, abundantly employing the power of being moved from themselves, they ran in a direction contrary [to their first course], and thus becoming most distant from their sources, they were at length ignorant that they were thence derived. (italics added to *audacity*)

> Mackenna's own translation of *Ennead*, V. 1 (Plotinus, *Enneads*, trans. Mackenna, vol. 4) reads:

> The evil that has overtaken them has its source in self-will, in the entry into the sphere of process, and in the primal differentiation with the desire for self ownership. They conceived a pleasure in this freedom and largely indulged their own motion; thus they were hurried down the wrong path and in the end, drifting further and further, they came to lose even the thought of their origin in the Divine.

39 Plotinus, "Preller-Ritter Extracts," trans. Mackenna, quotations on pp. 130, 131, referencing Plotinus, *Theologumena Arithmetica* and Plotinus, *Ennead*, V. 2.2. For

the Thomas Taylor translation of *Ennead*, V. 2.2, see *Collected Writings of Plotinus*, op. cit., p. 394: "the part which is in the plant . . . is most rash and insane." On "the Intellectual-Principle (the Divine-Mind) which is said, VI. 9,5, 'to sunder itself from The One in an act of self-assertion,'" see Plotinus, "Preller-Ritter Extracts," trans. Mackenna, quotation on p. 131. For the Thomas Taylor translation of *Ennead*, VI. 9.5, see *Collected Writings of Plotinus*, op. cit., p. 404: "intellect is not in itself dispersed, but is truly present with itself, and does not, in consequence of its proximity to *the One*, divulse itself, though in a certain respect *it dares to depart* from *the One*. . ." (italics added to *it dares to depart*). Mackenna's own translation of *Ennead*, VI. 9.5 (Plotinus, *Enneads*, trans. Mackenna, vol. 5) reads: "yet it is a principle which in some measure has dared secession."

40 Plotinus, "Preller-Ritter Extracts," ibid., p. 134, quotation on pp. 131–132.

41 Plotinus, "Preller-Ritter Extracts," ibid., p. 132.

42 Plotinus, "Preller-Ritter Extracts," ibid., pp. 132, 135, 136, 137.

43 Daniel Botkin, *Discordant Harmonies: A New Ecology for the Twenty-First Century* (New York: Oxford University Press, 1990), quotation on p. 25.

2 Christianity and Nature

On January 25, 1348, a major earthquake rocked the mountains of northern Italy.[1] During the main tremor, which lasted about two minutes, Giovanni da Parma was able to pray three Our Fathers and three Ave Marias. Fires, landslides, and floods swept the region. Buildings collapsed, villages were buried, people panicked, and hundreds were swept under the rubble. Two decades later, Italian scholar and humanist Francesco Petrarch (1304–1374), who was sitting in his study at the time, recorded his reactions:

> [The] vibrations arose so tremendously in large parts of Italy and Germany that a lot of people, who did not know about such tremors, thought the end of the world would be near. . . . The pavement trembled under my feet; when the books crashed into each other and fell down I was frightened and hurried to leave the room. Outside I saw the servants and many other people running anxiously to and fro. All faces were pale.[2]

In the Christian framework, natural disasters were often seen as punishments for human sin. Earthquakes, volcanoes, storms, droughts, famines, and plagues were retributions for human depravity. According to David Chester, many Old Testament passages characterize "the people who suffer [as] wicked and sinful, with God controlling alike the fates of people and nations." The Middle East and eastern Mediterranean region was:

> notable for its history of disasters, which include droughts, storms and floods, as well as earthquakes and volcanic activity. . . . Frequent and damaging earthquakes have occurred in the Holy Land. . . . The effects of volcanic activity, earthquakes and other natural disasters are used by the authors of the Hebrew *Bible* to support its dominant theodicy: that disasters represent punishment of human sinfulness by an often wrathful God.[3]

During the period in which Christianity rose to prominence as the dominant religion in Europe and on through the eighteenth century, "the explanation of major disasters that eclipsed all others was that these phenomena were either

Figure 2.1 1348 Earthquake. Fresco on the walls of St. Mary's Chapel in Karlstein Castle, near Prague depicting damage to Arnoldstein Castle in Austria from the 1348 earthquake.

Painted ca. 1361 by Nikolaus Wurmser.
Earthquake Archive, EERC Library, University of California, Berkeley, KZ7, detail. Courtesy of Jan Kozak Collection, NISEE-PEER, U.C. Berkeley.

44   *Autonomous Nature*

manifestations of divine power sent to punish human sinfulness and/or presaged the imminent end of the world."[4]

In this chapter, I argue that the Christian theologians, Augustine, Erigena, Aquinas, and Bonaventure, who introduced the two concepts—nature creating/nature naturing and nature created/nature natured—set up important questions for the Renaissance and Scientific Revolution, namely, what was the process of creation and how could nature as created world be understood, predicted, and controlled?

A major problem for the Christian era was how to explain the activity of nature in the created world. By accepting God as the initial Creator, the question still remained as to how changes took place in that world over time. What was the relation of the Creator to the created? If God as Being created the world *ex nihilo*, out of Not-Being, or nothingness, how was activity introduced and maintained? Once created, what was God's continuing, creating role? What was the explanation for sudden unusual events such as earthquakes, volcanoes, plagues, and droughts? Were they the results of God's retribution for human sin or nature's uncontrolled and unruly actions? And what were the implications of the gendering of God as male and nature (*Natura*) as female for subsequent philosophies of nature?

Developing during the Christian Middle Ages was a new framework about the nature of Nature that synthesized ideas from the Greco-Roman world with those of Christianity. During this period, the terms *natura creans/natura naturans* (nature creating) and *natura creata/natura naturata* (nature as created world) were introduced. The first concept referred to nature as an active, creating entity stemming from the very being of God; the second to the visible result of God's creation. For Christians, nature was not an eternally existing entity as some Greeks and Romans had believed, but was created by a single, uncreated God *ex nihilo*—out of nothing.

Augustine

Saint Augustine (354–430 C.E.) was the Bishop of the Roman province of Hippo in North Africa (present-day Algeria). During the century following Plotinus, he wrote the *Confessions* (composed 397–401 C.E.), the *City of God* (ca. 416 C.E.), and *On the Trinity* (ca. 428 C.E.), among other learned works. Early in life, he had studied the classics of Roman writers such as Virgil and at the age of nineteen was inspired to study philosophy after reading Cicero. He taught in Milan beginning in 384 C.E., where he was heavily influenced by Neoplatonism, especially Plotinus, through Bishop Ambrose of Milan. There, under the urging of his mother, he converted to Christianity in 386 C.E., later writing extensively about his conversion experience in the *Confessions*. He embraced the Council of Nicaea's concept of the Christian Trinity, or threefold nature of the Godhead. In 391 C.E. he was consecrated a priest in Hippo and ordained as Bishop in 395 C.E.[5]

Through Augustine, the Greco-Roman Neoplatonic philosophy merged with the Christian tradition, exerting a profound influence on subsequent Western culture. By combining the intellect's rational understanding and ability to make

Christianity and Nature   45

choices with the centrality and fallibility of the human will, the Greek intellectualist approach and the Christian voluntarist approach were brought together in both synthesis and conflict.

Foremost in Augustine's philosophy was the dualistic distinction between the Creator and the created world and between Creator and creature. Neoplatonism offered a philosophical framework that opposed ideas to matter and intellect to appearance, making possible an ascent to a reunification with God. But rather than positing a series of emanations from the One, as in Neoplatonism, Augustine emphasized the voluntarism of God's will in creating the natural world and of the human will in overcoming the temptations of the material world. For Christianity, evil is a moral issue, rather than a result of God's creative activity in the world. That distinction would be significant for the emerging dualism between God as an active, creating force and nature as created world.

In his book, *On the Trinity* (*De Trinitate*, ca. 428 C.E.), Augustine wrote about God as a "nature" (*natura*), or being, the defining characteristic of which was to exist eternally as an uncreated, single being, or One. In Book Fourteen, he stated that God, who is above humanity in terms of reason, consists of

> a nature (*natura*) not made, which made all other natures, great and small, and is without doubt more excellent than those which it has made, and therefore also than that of which we are speaking; viz. than the rational and intellectual nature, which is the mind of man, made after the image of Him who made it. And that nature, more excellent than the rest, is God.[6]

Thus God is a "nature" (or being) that is more excellent than anything he himself creates.

In Book Fifteen, Augustine characterized God as a "nature" that is "not created, but creates." Here he meant that God was an uncreated being, the nature, or defining characteristic of which, is to create. He wrote of God:

> If then we see something above this nature, and see it truly, then it is God, namely, a nature that is not created, but creates (*natura scilicet non creata, sed creatrix*). We must now show whether this is the Trinity—not only to believers by the authority of the divine Scriptures, but also, if we are able, to those who seek to understand by some kind of reason.[7]

God is therefore an uncreated, being above nature (or the created world) whose own nature, or essence, is to create the existing world out of nothing.

Augustine distinguished between humans and other animals, the former being superior to beasts by having a mind (*mens*) or "rational soul" (*animus*). The natural world comprised a hierarchy of (1) things that merely exist, (2) things that exist and live, and (3) things that exist, live, and understand. Humans possess understanding and can engage in reason and logic. The soul's capacity for understanding, reason, and illumination makes possible a return to and reunification with

46 *Autonomous Nature*

God. One can therefore move away from materiality and embrace the potential for eternal life. In later life, Augustine would emphasize God's grace, as opposed to the will alone, as a prerequisite for salvation.

Despite the influence of Plotinus on Augustine's idea of God as a creative force, a major difference existed between the two philosophers with respect to the roots of rebellion. Both thinkers provided a narrative of the fall as a descent from the original One/God—a falling away from the divine Source. For Plotinus, however, the fall resulted from the nature, or character of the world soul, whereas for Augustine it was a moral act, the result of choice. Plotinus described the fall of the Soul from the One as a rebellious audacity, or *tolma*, triggered by its willfulness and desire to assert itself by uniting with matter. First the Nous, or intellect, rebelled and separated from the One, followed by the World Soul, which had an upper and lower portion. Nature (*phusis*), as the lower part of the World Soul, thus had within itself the capacity for unruliness and lawlessness. This idea as an explanation for natural disasters would rise to prominence in the Renaissance.[8]

For Augustine, on the other hand, the fall likewise had two parts, but it was a moral fall, an act of choice. The first was the rebellion of the bad angels, one of which was Satan, who chose to reject God. The second was the sinful rebellion of Adam and Eve, responding to the temptation offered by Satan in the form of the serpent. Evil, for Augustine, was thus a result of human choice, rather than the rebellion of the World Soul as for Plotinus. Through the intellect, humans understand; through the will they act. The human soul is immaterial and immortal, but because it is also mutable, it can improve or deteriorate in accordance with the choices enacted by the will. People can turn away from God and toward the attractions of material things. They can thus choose to reject God and remain in a state of sinful depravation, or, alternatively, they can embrace good and with it the possibility of reunification with God. A narrative thus emerges of the human soul's immersion in the natural world, its subsequent liberation from material temptations, and, through faith in Christ, a reunification with God.[9]

Augustine's explanation of human sin as a choice dependent on the will set the tone for the emerging relationship between Christianity and nature. His idea of God as an active nature that is "uncreated, yet creating (*natura . . . non creata, sed creatrix*)" would be reflected, but modified, in the writings of Erigena, Aquinas, and Bonaventure. Yet in contrast to the Catholic Church's acceptance of Aquinas and Bonaventure, Erigena's writings about the creating power of God as one with Nature would be condemned as being too close to what in the eighteenth century became known as pantheism.

Erigena

The idea of God as "creating, but not created" became the center-point of Johannes Scotus (John the Scot) Erigena's (Eriugena, ca. 810–ca. 877 C.E.) five-volume work. Born in Ireland around 810 C.E., where he studied Greek, he moved to Laon, France (northeast of Paris) ca. 845 C.E., remaining there for thirty years. It was as head of the Palatine Academy that, during the years 864–866 C.E., he wrote the five books of the *Periphyseon* (from the Greek, *Peri Phuseos, Concerning Nature*). The *Periphyseon* was condemned in 1225 C.E. for what was later called pantheism. It was later translated in 1681 with the title, *The Division of Nature (De Divisione Naturae)*. Regarded as a Neoplatonist, Erigena was influenced both by Plato and Augustine and his work anticipated that of Spinoza and Hegel. His writings derived from Plotinus in the third century C.E. He also knew Augustine's works well and quoted from *On the Trinity (De Trinitate)* (although he differed from Augustine on the issue of dualism).[10]

Erigena's lack of orthodoxy, and the basis for its "pantheism" and condemnation by the Church in 1225 C.E., stemmed from three central positions. First, he maintained that God/Nature as an active Creator is one with the things He created and that God exists in all things, including humans. Second, the created world was not a creation out of nothing (*creatio ex nihilo*), but a flowing outward, or emanation, from the Creator. Third, he maintained that all creatures would be reunited with God at the end of creation, not just those predestined as the elect. These ideas infused his descriptions of the "creating process," the "created world," and "resurretion and return."[11]

The Creating Process

The unity of God/Nature (*Natura*) is Erigena's main subject. The first concern that Erigena's underlying system was pantheistic arose from his idea of the oneness of God and Nature. Nature is the totality of all things. In this original unity, Nature and God are one and the same thing and both are therefore divine. "God is the beginning, middle, and end of the created universe. God is that from which all things originate, that in which all things participate, and that to which all things eventually return."[12]

The second concern raised about Erigena's pantheist tendencies was that instead of the dichotomy between God and creation, basic to mainstream Christianity, Erigena developed a dialectical system. *Natura* was not only the universe itself, but also its processes of transformation. The original unity of God/Nature unfolded into the natural world as an out-flowing or emanation from the one God and then returned into a reunification of Nature/God. In its dialectical process of unfolding, he anticipated Hegel who, in fact, acknowledged Erigena's influence on his own work: "Philosophy properly speaking begin[s] in the ninth century with John Scottus Erigena."[13]

As a synthesis of Greek philosophy with Christianity, Erigena's God was both initial cause and final end. Here Neoplatonic and Christian accounts merged to explain the creating process. The process of development, however, was not cyclical as it was for many Greeks and Romans, but parabolic—one of creation, devolution, and return. He used the Biblical account of Genesis 2, or the fall of Adam and Eve, to explain the process. God/Nature exists at the outset, nature and human nature fall together with the ingestion of nature from the Tree of the Knowledge of Good and Evil, and are redeemed together through the Tree of Life (Christ). *Natura* (Nature/God) is a whole—both a totality and a process.

48   *Autonomous Nature*

Erigena's *Periphyseon* sets out his system of "nature creating" (*natura creans*), the "created world" (*natura creata*), and the "return" to God/Nature as four divisions. The first division (and first book) is that of God/Nature and deals with Nature (God) as actively creating, but not itself created (*natura creans et non creata*). The original unity simply exists and is therefore not created, but has the potential for creation within it. The second division (second book) concerns Nature that is both creating and created (*natura creans et creata*). Here Nature is active and in the process of creating the original ideas and primordial forms. The third division (third book) deals with the result of the creating process, the existing created world—the world perceived by and intelligible to human beings (*natura non creans et creata*). The fourth division (covered by books four and five) explicates the process of the return of the created world to God/Nature (*natura non creans et non creata*).[14]

The process of unfolding, as an emanation from God, meant a lack of demarcation between Creator and created. Out of the original unity of God/Nature emerged the Ideas or causes of the sensible world. These Ideas or pure Forms are contained within and emanate from the original unity, the Christian God. They are divine inasmuch as they stem from God, but they are also the causes of material things. From them are derived the minerals, plants, animals, and humans that constitute the known/knowable world of sensible beings. The divine is therefore contained within the created world, a position considered heresy by the Church.

The Created World

The created world comprises the sensible bodies perceived by humanity that result from the creating process. Yet these created bodies both participate in and are part of creating God/Nature, reinforcing the second concern over Erigena's pantheistic tendencies.

> It follows that we ought not to understand God and the creature as two things distinct from one another, but as one and the same. For both the creature, by subsisting, is in God; and God, by manifesting himself, in a marvelous and ineffable manner creates himself in the creature. . . .[15]

In the formation of the created world, the Divine plan can never be discerned by the human intellect, and humans cannot assign reasons to that which is beyond reason and surpasses human understanding. Quantities and qualities, which are initially incorporeal, come together to produce formless matter. From this matter arise the bodies of the sensible world, compounded out of the four elements. In them, quantities and qualities; forms and colors; length, width, and height; places and times, are manifest. But, "in sensible things no one can say how the incorporeal force of the seed, bursting out into visible species and forms, into various colors, into different fragrant odors, becomes manifest to the senses. . . . " Bodies receive quantities and qualities which are, as it were, "cloaked in garments" and seeds "burst forth through the numbers of places and times and the various species of animals, shrubs, and grasses." Yet humans cannot explain the reasons behind "the things which we see produced every year in the course of nature in the order of the seasons."[16]

The created world is metaphorically depicted as a river with smaller streams diverging from the central flow. Nature as an undivided whole contains invisible pockets within it, out of which develop the seeds of living things. The underlying force, or process, gives rise to life itself:

> There is a most general and common nature of all, created by the one First Principle of all; and from It, as from a very copious fountain, corporeal creatures, like rivulets, are channeled through hidden passages and break forth into the different forms of individual things. That force, coming forth through different seeds from the secret recesses of nature and first emerging in the seeds themselves, then mixed with the different fluids, bursts out into the individual, sensible species.[17]

The created world as an outward flow from the Creator raised deep concerns for the Church and Christianity, ultimately resulting in the condemnation of Erigena's system.

Resurrection and Return to God/Nature

The third problem for the Church and for the concerns about Erigena's pantheism was the idea that all creatures and all humans, not just a predestined few, would be reunited with God at the end of creation. Erigena devoted much of Book IV of the *Periphyseon* to an explanation of the dual animal and rational character of humankind and the process by which resurrection occurred. Sin comes from humanity's animal nature, but the divine allows for the return to God at the end of time. Reason and intelligence set humans apart from other living things. Each individual consists of an external, material body formed out of sensible matter and an inner, spiritual body consisting of sense, reason, intellect, and vital motion. The external body, like a cloak, changes in quality, quantity, size, and shape. The spiritual body, however, is immortal and unchanging—its form remains throughout the changes in the external garment. At the Fall of Adam, it was the material body that became subject to change and variation. At the return to Nature/God, the form, or unchanging spiritual body, is resurrected. The inner, spiritual human being is paradise—the tree of life planted in human nature. Christ as the tree of life is that inner spiritual being.[18]

At the end of time, all sensible things are reconstituted as Nature/God. Everything and everyone, whether good or evil, returns to the Nature that neither creates nor is created, a position at odds with the established Church. All sensible bodies, Erigena argued, go through a process of dissolution and return. Upon death, the human body dissolves into the four elements from which it was originally created. Death is thus the first step toward restoration. At the resurrection, the second step occurs, in which each individual receives back its own body and then, in the third step, is transformed into spirit. To understand how this is possible, Erigena gives

50   *Autonomous Nature*

examples, such as the way clouds originally formed from air can be resolved into pure air so that none of their original density remains and how that pure air is then consumed by the sun's brilliance. In the fourth step, the newly reformed body returns to its primordial causes. Then, in the fifth and final step, nature and its causes are reconstituted into a better state (paradise) as Nature/God.[19]

In that final state, nothing except Nature/God exists. Indeed, space and time are both part of and within the physical world and perish after the return. Moreover, after the resurrection the human form even lacks sex. Sexual differentiation occurred because of sin. Humanity returns to that form originally made in God's image. The process of resurrection of all beings and all humans, whether they have sinned or not, thus constituted the third major problem concerning Erigena's pantheistic framework.[20]

Erigena's *Periphyseon* was influential in the ninth century in France at Laon (where Erigena spent his last days) and at Corbie, and Auxerre. It was known to twelfth-century Neoplatonists through an edition by William of Malmesbury and circulated as a compendium entitled "The Key of Nature" (*Clavis Physicae*). It was also known to Aquinas and discussed by Hugh of St. Victor (ca. 1096–1141), Alain de Lille (ca. 1116/17–1202/03), Suger of Saint Denis (ca. 1081–13 January 1151), and philosophers at Chartres. It was condemned by Honorius III in 1225 and by Gregory XII in 1585, but nevertheless influenced Meister Eckhart (ca.1260–ca.1328) and Nicholas of Cusa (1401–1464). During the Scientific Revolution, it was published in Germany in 1638, in England in 1681, and was placed on the Catholic index in 1685 where it remained until the 1960s.[21]

Both Erigena and Augustine were thus deeply influenced by Neoplatonism and particularly by Plotinus. Both contributed in different ways to the concept of God as nature creating/nature naturing. During the thirteenth century, Thomas Aquinas would add to this emerging framework by drawing on the works of Aristotle.

Aquinas

While Augustine and Erigena had been influenced by Plotinus, Thomas Aquinas (1225–1274) owed much of his approach to Aristotle and sought to reconcile the latter with Christianity. Aquinas was a member of the Dominican Order and received his degree at the University of Paris where he taught for several years. Latin translations of Aristotle's works, including the *Physics, Metaphysics, On the Heavens*, and *On Generation and Corruption*, had been introduced into Europe in the twelfth century, and Aquinas was exposed to them early in his schooling. His voluminous works included numerous commentaries on Aristotle's writings. He was canonized as a Saint by Pope John XXII in 1323.[22]

While Erigena had used the terms *natura creans* and *creata*, Aquinas began a tradition of using *natura naturans* to which was added *natura naturata* by Bonaventure. The two Latin terms (*natura naturans* and *naturata*) seem to have been introduced by

> translators into Latin of the *Commentary* of Averroes on the *Physics* of Aristotle, which would make the terminology to be of specifically Latin origin dating from the early 1200s. . . . The translator uses the expression *natura*

naturata and introduces into Latin the verb '*naturare*' under the form *natura-tur* as a rendering of Aristotle's expression [*phusis*]. Given the verb form, the formation of the present participle form *naturans* to contrast with the past participle *naturata* becomes an inevitable natural language formation. . . . Yet *naturare* itself as a new Latin verb, at least in the translation of Averroes' *Commentary on Aristotle's Physics*, correlates directly with Aristotle's Greek.[23]

In his *Summa Theologica* (written 1265–1274), Aquinas discussed the concepts of God and Nature. He was aware of Augustine's idea of God as *natura creata* and *creatrix*, but used the term *natura naturans* as an active power permeating all of nature:

> 'All nature' (*Natura vero universalis*) refers to an active power existing in some universal principle of nature (*universali principio naturae*), in some heavenly body, for instance, or in some superior substance, in the way in which some call God *natura naturans*. Such a force intends the perfection and preservation of the entire universe, and this requires the interplay of generation and corruption among particular beings. From this point of view, then, the corruption and defects in certain things are natural; they are attendant not upon the tendency of form, the principle of existence and perfection, but upon a tendency in matter, which is proportionately conjoined to a form according to the determination of the universal agent.[24]

For Aquinas, matter rather than form is corruptible. Disorder is introduced into nature by sin, via the human will. Nevertheless, he argued that this disorder lessens neither the good of nature nor that of human nature because the form (the rational soul in humans), which is based on perfection, remains incorruptible. It is matter that is subject to corruption and deficiency. Natural disasters occur because of the corruptibility of the material world which is made up of contrary elements. Sin in and of itself does not do permanent damage to the good of nature.[25]

Among Aquinas's early writings was a treatise "On the Principles of Nature" (*De Principiis Naturae*, 1252–1256). This treatise, an explication of Aristotle's *Physics*, formed the basis for his views on generation and corruption and hence the basis for disorder and disasters in the natural world.[26] For Aristotle, the form and matter were united in any given being, rather than the form being part of a separate world of unchanging ideas as for Plato and Plotinus. Following Aristotle, Aquinas wrote:

> Matter is said to be the cause of form insofar as form exists only in matter, and similarly form is the cause of matter in that matter has actual being only through form; matter and form are correlatives, as is said in *Physics 2*.[27]

There were four causes of any particular individual—material, efficient, formal, and final. Aquinas went on to explain that the matter and form are said to be intrinsic to the thing because they are parts that constitute the thing. Conversely, the efficient and final causes, he stated, are said to be extrinsic, because they are outside the thing.[28] Change was explained through generation and corruption.

52 *Autonomous Nature*

Generation was the process by which the union of matter and form took place. Generation, however, could occur only through an agent, or efficient cause, which gave motion to matter. In natural things, the agent (efficient cause) was in the natural entity. An acorn became a tree; a child became a man. In artificial things the agent was outside the entity, as when clay was molded into a bowl by a potter. Through the fusion of matter and form arose an actual individual. The form or soul of a particular body constituted its essence. A human being was a rational animal by fusion of its body with a rational soul.[29]

Corruption in the material world accounted for natural disasters. Whereas moral evils occurred in humans as free moral agents, natural evils were the result of contrary actions within prime matter, owing to matter's corruptibility and deficiencies. Joseph Magee writes that

> Natural evils include not only people and animals being in the wrong place at the wrong time (e.g. when a hurricane is obeying the laws of nature), but also the more pervasive evils of death and disease. . . . Aquinas does not consider natural evils to result only from things external to the victim of natural evils obeying natural laws. Rather, for Aquinas, natural evils include natural disasters, but also result from the very nature of things as material.[30]

Not only did Aquinas contribute to an explanation for natural disasters, he also contributed to the concept of natural law.

Natural Law

Human reason is the basis for Aquinas's concept of natural law, a theory that would become important during the Scientific Revolution and Enlightenment. But what does natural law mean to Aquinas and how does it relate to *natura naturans* or active nature? Could the rationality and logic of natural law be used to understand human nature and human society and, ultimately, in conjunction with the laws of nature, to contribute to an orderly world? By the late seventeenth century, the concepts of natural law and the laws of nature would indeed help to conceptualize a rationally ordered society and a rationally ordered cosmos in which the unpredictability of nature could be managed and controlled. The idea of natural law as stemming from God's eternal law can be found in the thought of Aquinas.

In the section *On Law and Natural Law* (1271 C.E.) in the *Summa Theologica*, Aquinas sets up eternal law, existing in God, as the foundation of natural law. Natural law derives from God's eternal law or rational plan that orders all of nature. Through their ability to reason, humans receive natural law from God and can understand both its principles and the moral code underlying all of reality. They are morally bound by natural law. Moreover, it is part of human nature to do good deeds.[31]

God governs the world through eternal law, or divine reason. Everything created by him is subject to his eternal law. He commands the whole of nature. God's intrinsic active principle is impressed on all natural things. "The impression of the intrinsic active principle in natural things is like the promulgation of the law to men," Aquinas wrote. Human laws, however, pertain only to rational beings, i.e., other humans. Irrational beings, such as other animals, cannot be the subject of human laws. They nevertheless participate in the divine reason through obedience to it.[32]

Although all things are perfectly ordered by eternal law, no humans can fully comprehend it. God inserts that law into human minds. Natural law is thus "a participation in us of the eternal law." Acting through reason and will, rather than by natural appetite as in irrational creatures, humans create human laws. Human law, therefore, derives from natural law.[33]

> Under God the lawgiver different creatures have different natural inclinations, such that something that is according to the law for one, is against the law for another, as if I should say that to be ferocious is in some way the law of the dog, but would be against the law of the sheep or any other mild animal. There is then the law of man, which arises from the divine ordinance according to his proper condition, that he should act according to reason.[34]

Sensuality in humans is a deviation from the law of reason. Humans turned away from God and became more like "beasts." In other animals, sensuality is a direct inclination directed toward the preservation of the species or the individual and hence is part of the common good. In humans, it is also directed toward the common good as long as it is subject to reason, but if reason is rejected, humans abandon their own dignity and become more like the lower animals.[35]

Natural law is thus a reflection of the order of nature as set out by God and governed by divine reason. By participation in reason, humans have some limited power to understand and hence to use the laws of nature for beneficial purposes. Other animals participate in the order of nature, but not in reason. It is through reason, therefore, that humans can understand *natura naturans* as an active, creating power and potentially to gain some control over the things of nature. To Aquinas's use of *natura naturans* as an active creating power, Bonaventure would add the concept of *natura naturata*, in a further synthesis of Plotinus and Aristotle with Christianity.

Bonaventure

Franciscan Cardinal, Saint Bonaventure (1217–1274), lived and wrote at the same time as Thomas Aquinas (1225–1274). He was an Italian theologian, believed by some to be the most brilliant and astute of all the medieval Catholic theologians. In 1243 at the age of twenty-six, he entered the Franciscan order and ten years later was awarded the Franciscan Chair at the University of Paris. In 1255 he became a master (equivalent of a doctor). He was eight years younger than Aquinas, but owing to an irregularity, the two were actually awarded their degrees during the same year in 1257. He was named a Cardinal Bishop by Pope Gregory X and two centuries after his death was canonized as a Saint by Pope Sixtus IV in 1482.[36]

54   *Autonomous Nature*

Heavily influenced by Augustine and Dionysius the Areopagite, Bonaventure, like Aquinas, drew on the philosophy of Aristotle, but in contrast to Aristotle argued for a creation out of nothingness, *creatio ex nihilo*, rather than an eternally existing world. He also read and was affected by Plato and Plotinus as well as by mysticism. His works were collected in twelve volumes, the first four of which were entitled, *Commentaries on the Four Books of Sentences of Master Peter Lombard* (written at the age of twenty-seven) along with eight additional volumes containing shorter but highly significant works, including the *Breviloquium* (vol. 9).

Bonaventure seems to have been the first of the Christian theologians to use both *natura naturans* and *natura naturata*. In his *Commentaries on . . . Peter Lombard*, he used both phrases as well as *natura creata* in the same sentence.[37] Here, in speaking about God the Son, he wrote that nature was not to be taken generally, but instead as created nature (*natura creata*). The Son is not above the eternal generation, or nature naturing (*natura naturans*), meaning God, but is nevertheless above created nature (*naturam creatam*), i.e., the ordinary nature natured (*natura naturata*).[38]

As a scholar of Augustine, he was presumably aware of Augustine's use of *natura non creata, sed creatrix*. As a contemporary of Aquinas he was probably also aware of Aquinas's use of *natura naturans*. He did not use Erigena's term, *natura creans*, however. The footnote to the Latin edition of Bonaventure's *Commentaries on . . . Peter Lombard* cites Augustine as using *natura creata* as well as Erigena's *de Divisione Naturae*, but those references would seem to have been added by his nineteenth-century editors.[39]

In the section "On Creation" in his *Breviloquium*, he wrote: "The entire fabric of the universe was brought into existence in time and out of nothingness, by one first Principle, single and supreme, whose power, though immeasurable, has disposed all things by measure and number and weight." By using the words "in time and out of nothingness," he distanced himself from Aristotle's concept of an eternal, uncreated cosmos, as well as the pantheistic concept of "an eternal material principle."[40]

Whereas Plotinus based his philosophy on the One, Bonaventure rooted his in the personal God of *Genesis*. Together the "One First Principle" and the "Trinity of God" constituted the foundation of creation. Following Plotinus, Bonaventure drew on the concept of emanation to explain the way in which the Son and Holy Spirit emerged from within God the Father, but unlike Erigena refrained from applying it to the created world. Within the undivided Supreme Principle there were two modes of emanation, one through nature and one through the will. The first mode of emanation through nature produced the Son by the process of generation. The second mode of emanation through the will produced the Holy Spirit, by a process he called spiration-procession. The First person was uncreated, the Second originated from the first by generation, and the Third from the first and second though "spiration or procession."[41]

As a Trinity, God therefore consisted of three persons, the Father, Son, and Holy Spirit. The *Bible* establishes that God has an "Offspring whom He supremely

Christianity and Nature 55

loves" that was made flesh, i.e., Christ the Son. Bonaventure referred to and quoted Augustine on the nature of the Trinity:

> The proof of God's existence is founded not only upon the authority of the divine books, but also upon the entire natural universe around us, to which we ourselves belong, and which proclaims that it has a transcendent Creator: a Creator who granted us natural intelligence and reason. . . ."

The Creator was above the creation, alive, immortal, and omnipotent. The Trinity comprised a "begetting Mind, a begotten Word, and a Love that unites them."[42]

Within the original Oneness existed the order of nature, the order of wisdom, and the order of goodness. God operates according to the idea that "in the beginning before time was, the luminous, translucent, and opaque natures were brought from non-being into being," in accordance with the words of Genesis I that "in the beginning God created the heavens and earth." The heavens were the luminous nature, the waters, the translucent nature, and the earth the opaque nature. Creation did not occur in "utter chaos," but in three levels of perfection—high, middle, and low. The heavens on high were perfect, the intermediate nature was not yet separate, and the lowest level, the earth, was still unorganized. The complete separation took three days.[43]

In describing the hierarchies of the resulting creation, Bonaventure followed a combination of Christian and Aristotelian frameworks. There were three heavens, the empyrean heaven (which was the motionless abode of God), the crystalline sphere (which accounted for the daily motion of the heavens), and the firmament, or sphere of the fixed stars. Below these were the spheres of the seven known planets: Saturn, Jupiter, Mars, the Sun, Venus, Mercury, and the Moon. Below the Moon were the spheres of the four elements: Fire, Air, Water, and Earth which mixed and combined to form the corruptible world. The elements existed in active and passive opposition to each other through which activity took place on the earth. The three heavenly spheres, seven planetary spheres, and four elemental spheres constituted a "universe so beautiful in its proportions, so complete and orderly, that in its own way it offers an image of its Principle."[44]

Consistent with Platonism (and in contrast to Aristotelianism), Ideas, for Bonaventure, existed in the mind of God, not in the things of nature. Universal forms in the mind of God were models for material things. Matter was pure potentiality shaped by ideal forms. Animals were material beings that existed as mere shadows of ideal forms. Humans as intellectual beings were links between the material and intellectual worlds. Through the intellect humans could grasp the perfection and grace of God.

Although Bonaventure followed Plotinus in employing the process of emanation to explain the existence of the Trinity, he did not hypothesize a rebellious aspect to that process that could result in natural disasters. Thus while God, like the One, is eternal, perfect, and the creator of the natural world, the creating process does not produce a rebellious form of Nature. Instead, consistent with the

56 *Autonomous Nature*

Bible, rebellion enters the world through the disobedience of Eve. Because "man" was made out of nothingness, he was imperfect by nature and "had the capacity of acting for ends other than God; of acting for himself instead of for God. . . . That precisely is sin: the vitiation of mode, species, and order." Sin is thus a defect and a corruption. It is not a desire for evil, but a falling away from good. Yet Bonaventure draws on Plotinus to explain how corruption enters humanity through the will. "[F]ree will, by falling away from the true Good, corrupts its own mode, species, and order; hence, sin as such proceeds from the will . . . and resides in the will." It "occurs whenever the will, through fallibility, mutability, and indifference, spurns the indefectible and immutable Good and cleaves to the mutable."[45]

God, as the first Principle, was not only "utterly powerful in the act of creating," but also "utterly just in the act of governing." He allowed Eve to be tempted so that humans could learn the "merit of obedience and the evil of rebellion." The serpent's strategy was to overcome the weaker of the two sexes and to wrest "consent from the free will, a faculty of both intellect and will" by taking on the power of the human appetite. He appealed both to Eve's powers of reasoning and to her bodily desire to taste the fruit of the tree.[46]

Bonaventure thus drew on both Plotinus and Aristotle to formulate a Christian philosophy of rationality and faith that explained God as a creating being and the natural world as a created entity. Bonaventure's terms *natura naturans* and *natura naturata* and Erigena's concept of God/Nature would be merged by subsequent philosophers into a pantheistic monism uniting the creating and created worlds.

Other scholastics who used the terms *natura naturans* and/or *natura naturata* included Vincent de Beauvais (1264), Raymond Lull (1315), and Peter of Abano (1315). In the seventeenth century Giordano Bruno (1600), John of St. Thomas (1589–1644), Bernard Sannig (1685), and Francis Bacon also used the terms. Spinoza developed the concepts into a major tenet of his philosophy, while also drawing on Aquinas's idea of natural law to argue for the rationality of both nature's law and human law as models for the cosmos and society.[47]

Conclusion

By the fourteenth century C.E., a number of ideas had appeared that would ultimately influence the idea of controlling nature. First and foremost was the concept of God as an active, creating force. For Augustine, God was a "nature that is not created, but creates (*natura . . . non creata, sed creatrix*)." For Erigena, God/Nature was creating, but not created (*natura creans et non creata*). The latter introduced a fourfold framework of *natura creans* and *natura creata*: (1) out of God/Nature, by means of (2) the creating process (*natura creans*), (3) flowed the sensible world (*natura creata*) much like streams emerging from a central fountain, (4) to be reunited with Nature/God at the resurrection.

For Aquinas, God was an eternal, active, creating power, *Natura naturans*, the divine plan of the world. From this intrinsic, active principle came natural law, which could be rationally understood by human beings and to which they were morally bound. Rationality, natural law, and morality were integrally related. For

Bonaventure who used both *natura naturans* and *natura naturata*, God was utterly powerful in the act of creating and utterly just in the act of governing the created world.

By the late Middle Ages, explanations of natural disasters such as the 1348 earthquake, began to move beyond the idea of God's retribution and toward the acceptance of natural disasters as part of everyday life. Although some commentators, such as Konrad von Megenberg in his *Buch der Natur* (written in 1349), drew on both Aristotelian and scholastic rationales, Christian Rohr argues that there was no consistent belief that earthquakes and other disasters were divine punishments, inasmuch as the righteous and the sinful both perished. People could be viewed, instead, as living within the vicissitudes of an autonomous nature, ultimately recovering from devastating events. Meanings began to move toward the type of rational explanation that would come to characterize early modern science, one in which "[M]an perceives nature as wild and unpredictable," states Rohr, "but lives and copes with it by using technical knowledge."[48]

Together, the Christian theologians contributed to the idea of a cosmos created *ex nihilo* by a rational God—one that could be understood through the laws of nature—and to a rational human society based on natural law. Upsetting this trajectory, however, intervened the Renaissance dichotomy between a personified Nature acting as God's instrument *versus* a recalcitrant Nature acting *not* in accordance with God's plan, but on "her" own, creating the problem of how "she" could be managed. By the time of the seventeenth-century Scientific Revolution, natural philosophers would respond by developing the idea of controlling a free, recalcitrant nature through the laws of nature and a disorderly society through natural law.

Notes

1 Christian Rohr, "Man and Natural Disaster in the Late Middle Ages: The Earthquake in Carinthia and Northern Italy on 25 January 1348 and Its Perception." *Environment and History* 9 (2003): 127–149, see p. 132.
2 Quoted in Rohr, "Man and Natural Disaster," p. 134.
3 David K. Chester and A. M. Duncan, "Volcanoes, Earthquakes, and God: Christian Perspectives on Natural Disasters," *SECED Newsletter* (The Society for Earthquake and Civil Engineering Dynamics) 22, no. 4 (2011): 1–6, quotations on p. 1, 2; David Chester, "The Theodicy of Natural Disasters," *Scottish Journal of Theology* 51, no. 4 (1998): 485–505.
4 Chester and Duncan, ibid., quotation on p. 2.
5 On Augustine's life and philosophy, see Michael Mendelson, "Saint Augustine," *The Stanford Encyclopedia of Philosophy*, http://plato.stanford.edu/entries/augustine/ (2010) and "Augustine of Hippo," http://en.wikipedia.org/wiki/Augustine_of_Hippo.
6 Augustine, *On the Trinity* [*De Trinitate* (ca. 428 C.E.)], Books 8–15, ed. Gareth B. Matthews, trans. Stephen McKenna (Cambridge, UK: Cambridge University Press, 2002), Bk. 14, Ch. 16. Latin: http://www.thelatinlibrary.com/augustine/trin14.shtml:

> Est igitur natura non facta quae fecit omnes ceteras magnas paruasque naturas eis quas fecit sine dubitatione praestantior, ac per hoc hac etiam de qua loquimur rationali et intellectuali quae hominis mens est ad eius qui eam fecit imaginem facta. Illa autem ceteris natura praestantior deus est.

58   *Autonomous Nature*

7 Augustine, *On the Trinity*, Bk. 15, Ch. 1, pp. 167–168, quotation on p. 168. Latin: Augustini De Trinitate Liber XV: http://www.thelatinlibrary.com/augustine/trin15.shtml:

> Supra haec ergo naturam si quaerimus aliquid et uerum quaerimus, deus est, *natura scilicet non creata, sed creatrix*. Quae utrum sit trinitas non solum credentibus diuinae scripturae auctoritate, uerum etiam intellegentibus aliqua si possumus ratione iam demonstrare debemus. Cur autem "si possumus" dixerim res ipsa cum quaeri disputando coeperit melius indicabit.

> *Natura creatrix* refers to nature as a creatress.

8 Plotinus, *Enneads*, in *Collected Writings of Plotinus*, trans. Thomas Taylor (Frome, Somerset, UK: Prometheus Trust, 1994), "On Nature, Contemplation and the One," *Ennead*, III, 8, pp. 123–134; John Deck, *Nature, Contemplation, and the One: A Study in the Philosophy of Plotinus* (Toronto: University of Toronto Press, 1967), Ch. 6, pp. 64–72, esp. p. 64.

9 On the fallen angels see, Augustine, *Confessions*, http://sparks.eserver.org/books/augustineconfess.pdf, Bk. 1, sec. 10: "For in more ways than one do men sacrifice to the rebellious angels"; Bk. 7, sec. 59:

> If the devil were the author, whence is that same devil? And if he also by his own perverse will, of a good angel became a devil, whence, again, came in him that evil will whereby he became a devil, seeing the whole nature of angels was made by that most good Creator?

> See also *The Confessions of Saint Augustine*, ed. J.J. O'Donnell (Oxford, UK: Oxford University Press, 1992, electronic edition), http://www.stoa.org/hippo/noframe_entry.html: "1.17.27, 'non enim uno modo sacrificatur transgressoribus angelis'; for the fall of the angels, cf. 7.3.5, 'ex bono angelo diabolus factus est'; see [*De Civitate Dei, The City of God*], 12.1–9, and for the Christian identification of the fallen angels as *daemones*, [*De Civ.*], 9.19."

10 On Irish theologian, Johannes (John) Scotus (Scottus) Erigena (Eriugena), also known as John the Scot, ca. 810–ca. 877 C.E. (not to be confused with John Duns Scotus, 1266–1308 C.E.), see http://www.britannica.com/EBchecked/topic/191466/John-Scotus-Erigena; Johannes Scotus Eriugena, *Periphyseon* (*De Divisione Naturae*), *English and Latin*, 9th c., ed. I.P. Sheldon-Williams, with the collaboration of Ludwig Bieler (Dublin: Institute for Advanced Studies, 1968–). *Periphyseon* (*Concerning Nature*) derives from the Greek term *Peri Phuseos* (*physis*). For the philosophical influences on Erigena, his use of a variety of Biblical translations, and a summary and translation of the *Periphyseon*, see Joannes Scotus Eriugena, *Periphyseon: On the Division of Nature*, trans. Myra L. Uhlfelder with summaries by Jean A. Potter (Indianapolis, IN: Bobbs-Merrill, 1976). On Erigena's method, narrative, and anthropology, see Willemien Otten, "The Dialectic of the Return in Eriugena's *Periphyseon*," *Harvard Theological Review* 84, no. 4 (1991): 391–421 and idem, *The Anthropology of Johannes Scottus Eriugena* (Leiden: E.J. Brill, 1991). The first printed edition of Erigena's *Periphyseon* was by Thomas Gale in 1681 under the title *De Divisione Naturae* (*The Division of Nature*) (Oxford: Thomas Gale, 1681). Erigena knew Augustine's works well and quoted from *De Trinitate*. See Hilary Anne-Marie Mooney, *Theophany: The Appearing of God According to the Writings of Johannes Scotus Eriugena* (Tubingen, Germany: Mohr Siebeck, 2009), Ch. 3, "God Appears in Creation," p. 97.

11 Eriugena, *Periphyseon* (*De Divisione Naturae*), ed. Sheldon-Williams, ibid., Bk. I, Ch. 72, p. 209:

> So when we hear that God makes all things we ought to understand that God is in all things, that is, that He is the Essence of all things. For only He truly exists by Himself, and He alone is everything which in the things that are is truly said to be.

Latin, ibid., p. 208, lines 581A 18–22:

> Cum ergo audimus Deum omnia facere nil aliud debemus intelligere quam Deum in omnibus esse, hoc est, essentiam omnium subsistere. Ipse enim solus per se vere est, et omne quod vere in his quae sunt dicitur esse ipse solus est.

Periphyseon, ibid., Bk. I, Ch. 3, p. 39: "He is the Essence of all things Who alone truly is." Latin, ibid., p. 38, lines 443B 25–26: "Ipse [Deus] nanque omnium essentia est, qui solus vere est"; Bk. III, Ch. 23, p. 183:

> [It] is more or less agreed between us that *all things are from God and that God is in all things* and that they were made from nowhere but from Him—since from Him and through Him and in Him all things are made. . . .

Latin, ibid., p. 182, lines 688A, 26–31:

> Nos autem nostrae rationcinationis iter teneamus, et quoniam prope modum inter nos est *confectum onmia ex deo et deum in omnibus esse* et non aliunde nisi ex ipso facta esse—quoniam ex ipso et per ipsum et in ipso facta sunt omnia—. . . .

See also William Turner, History of Philosophy, Jacques Maritain Center, Ch. XXV, "John Scotus Erigena," http://www3.nd.edu/Departments/Maritain/etext/hop25.htm. For an analysis of the history and reasons for and against Erigena's alleged pantheism, see Dermot Moran, *The Philosophy of John Scottus Eriugena: A Study of Idealism in the Middle Ages* (Cambridge, UK: Cambridge University Press, 1989), pp. 84–89; idem, "Pantheism in John Scottus Eriugena and Nicholas of Cusa," *American Catholic Philosophical Quarterly* (Winter 1990): 131–152.

12 Eriugena, *Periphyseon*, Bk. III, lines 621a–622a, quoted in http://www.enlightened-spirituality.org/John_Scottus_Eriugena.html. See also *Periphyseon*, ed. Sheldon-Williams, op. cit., Bk. III, p. 33: "He is the causal Beginning of all those things, and the essential Middle which fulfils (them), and the End in which they are consummated and which brings to rest every motion and imposes tranquility. . . . " Latin, ibid., p. 32, lines 622A, 17–21: " . . . et super omnia quae dicuntur et intelliguntur et omni sensu percipiuntur dum sit horum omnium principium causale et medium implens essentiale et finis consummans omnemque motum stabilitans quietumque faciens et ambitus omnia quae sunt et quae non sunt circumscribens."

13 Otten, "Dialectic of Return in Eriugena's *Periphyseon*," op. cit., pp. 399–405, 408, 411; G.W.F. Hegel, *Lectures on the History of Philosophy, 1825–6*, Vol. III, *Medieval and Modern Philosophy*, rev. ed., trans. and ed. George F. Brown (New York: Oxford University Press, 2009), p. 42.

14 Otten, "Dialectic of Return in Eriugena's *Periphyseon*," op. cit., pp. 399–400; http://www.britannica.com/EBchecked/topic/191466/John-Scotus-Erigena.

15 Eriugena, *Periphyseon*, ed. Sheldon-Williams, op. cit., Bk. III, pp. 161, 163; Latin, ibid., p. 162, 678C, lines 1–4: "Nam et creatura in deo est subsistens et deus in creatura mirabili et ineffabili modo creatur se ipsum manifestans, invisibilis visibilem se faciens et incomprehensibilis comprehensibilem et occultus apertum et incognitus cognitum. . . . "

16 Eriugena, *Periphyseon: On the Division of Nature*, trans. Uhlfelder, op. cit., Bk. III, Ch. 14, pp. 177–189, see esp., pp. 184, 177–178, quotations on pp. 188, 181, 186; Eriugena, *Periphyseon*, ed. Sheldon-Williams, Bk. III, Ch. 14, pp. 127–135, quotations on pp. 147 [Latin, p. 146, 671D,1–6]; 133[Latin, p. 132, 665C, 27]; 141 [Latin, p. 140, 669C, 30–34].

17 Eriugena, *Periphyseon*, trans. Uhlfelder, Bk. IV, pp. 216–217; Eriugena, *Periphyseon*, ed. Sheldon-Williams, Bk. IV, p. 19 [Latin, p. 18, 750A, 31–37].

18 Eriugena, *Periphyseon*, trans. Uhlfelder, pp. 217, 259–260, 263, 266; Eriugena, *Periphyseon*, ed. Sheldon-Williams, Bk. IV, p. 21 [Latin, p. 20, 250B, 10–15]; p. 109 [Latin,

60 *Autonomous Nature*

p. 108, 286C, 11–14]; p. 141 [Latin, p. 140, 801A, 15–17]; pp. 193, 195 [Latin, p. 192, 823B; 194, 824C, 25–31].

19 Eriugena, *Periphyseon*, trans. Uhlfelder, pp. 209, 287; Eriugena, *Periphyseon*, ed. Sheldon-Williams, Bk. IV, p. 5 [Latin, p. 4, 743C, 28–30]; [Note: Bk. V, ed. Sheldon-Williams, has not been published].

20 Eriugena, *Periphyseon*, trans. Uhlfelder, pp. 287, 291, 293–294, 299.

21 "John Scottus Eriguena," *Stanford Encyclopedia of Philosophy*, http://plato.stanford.edu/entries/scottus-eriugena/ (2004); "Erigena," *Chamber's Encyclopedia: A Dictionary of Universal Knowledge* (Philadelphia: Lippincott, 1889), Vol. 4, p. 413.

22 For background on Thomas Aquinas's life and theology, see "Thomas Aquinas," http://en.wikipedia.org/wiki/Thomas_Aquinas and "Saint Thomas Aquinas," http://www.themiddleages.net/people/aquinas.html.

23 John Deely, *Four Ages of Understanding: The First Postmodern Survey of Philosophy from Ancient Times to the Turn of the Twenty-First Century* (Toronto: University of Toronto Press, 2001), see pp. 135–140, quotation on pp. 139–140. See Deely, p. 139, note 139:

> The text cited by [Hermann] Siebeck, ["Uber die Entstehung der Termini *natura naturans* und *natura naturata*," *Archive für Geschichte der Philosophie*, 1890: 370–378, on p.], 374, and by [Henry] Lucks, ["*Natura Naturans—Natura Naturata*," *The New Scholasticism* 9, no. 1 (1935): 1–24]: 14 following him, is from Averroes, *Comment. ad Aristotl. Phys.* II, I, 11: "*Necesse enim est, ut initium medicinandi sit ex medicina et non inducit ad medicinam, et non est talis dispositio naturae apud naturam; sed naturatum ab aliquo ad aliquid venit, et naturatur aliquid; ipsum igitur naturari aliquid non est illud ex quo incipit sed illud ad quod venit. . . . Hoc igitur nomen natura derivatur a nomine eius quod advenit sive quum dicimus ipsum esse naturatum; et hoc intendebat quum dixit: 'sed naturatum' etc., i.e. sed naturatum ab illo a quo generatur ad aliquid venit, et dicitur ipsum naturari aliquid.*"

> See also Deely, p. 139, note 141, citing Olga Weijers, "Contribution à l'histoire des termes '*nautura naturans*' et '*natura naturata*' jusqu'à Spinoza," *Vivarium* 16, no. 1 (1978): 70–80:

> It may be that Michael Scot (fl. 1200) created the active form *natura naturans*. At least, one first finds the expression within his *Liber introductorius* (in a context relative to the hope of engendering and to the power of God to modify natural infertility): *cum Deus sit natura naturans et ideo superet naturam naturatam* [since God is 'nature naturing' and therefore overcomes 'nature natured.'].

24 Thomas Aquinas, *Summa Theologica*, Latin text and English translation (London: Blackfriars, 1963 [1265–1274]), vol. 26, IaIIae, 85, 6, reply, p. 101 (Latin, p. 100, note 9, cf. Augustine, *De Trinitate* [*On the Trinity*], Bk. 15, I, PL. 42, 1057, uses *natura creata* and *natura creatrix*); Latin text, S. Thomae Aquinatis, *Summa Theologica* (Rome: Marietti, 1952), vol. 1, Pars IaIIae, q. 85, a. 6, p. 391.

25 Aquinas, *Summa Theologica*, Latin text and English translation, vol. 26, IaIIae, q. 85, "The Damage to the Good of Nature," p. 79; Latin: "Videtur quod peccatum non diminuat bonum naturae," p. 78.

26 Thomas Aquinas, *Selected Writings*, ed. and trans. Ralph McInerny (New York: Penguin, 1998), "On the Principles of Nature," pp. 18–29.

27 Aquinas, *Selected Writings*, "On the Principles of Nature," quotation on p. 25.

28 Aquinas, *Selected Writings*, "On the Principles of Nature," p. 23, Ch. 3.

29 Aquinas, *Selected Writings*, "On the Principles of Nature," p. 22.

30 Joseph M. Magee, "Thomistic Philosophy Page," http://www.aquinasonline.com/Topics/probevil.html.

31 Aquinas, *Summa Theologica* (*ST*), in *Selected Writings*, "On Law and Natural Law," IaIIae 90–94, pp. 611–652; Latin text, Aquinatis, *Summa Theologica* (1952), vol. 1, Pars IaIIae, q. 90–94, pp. 410–430.

Christianity and Nature 61

32 Aquinas, *Summa Theologica* (*ST*), in *Selected Writings*, "On Law and Natural Law," *ST* IaIIae 93, 4; quotation, 93, 5, response, Ad 1.

33 Aquinas, *Summa Theologica* (*ST*), in *Selected Writings*, "On Law and Natural Law," *ST* IaIIae 91, 2; 93, 2, 3; 91, 3; 90, 4; 91, 4; 91, 2; 91, 3; quotation 91, 4.

34 Aquinas, *Summa Theologica* (*ST*), in *Selected Writings*, "On Law and Natural Law," *ST* IaIIae 91, 6, response.

35 Aquinas, *Summa Theologica* (*ST*), in *Selected Writings*, "On Law and Natural Law," *ST* IaIIae 91, 6, response.

36 On Bonaventure's life and work, see Tim Noone and R.E. Houser, "Saint Bonaventure," http://plato.stanford.edu/entries/bonaventure/ and "Bonaventure," http://en.wikipedia.org/wiki/Bonaventure.

37 Bonventure, Saint, *Opera omnia . . . edita studio et cura PP Collegii a S. Bonaventura, ad plurimos codices mss. emendata, anecdotis aucta, prolegomenis scholiis notisque illustrata*, 11 vols. (Ad Claras Aquas, Quarracchi, Ex typographia Collegii S. Bonaventurae, 1882–1902), vol. 3, p. 197 [p. 218, online] (Distinctio VIII, Article II, Question III), Dubio, Dubia Circa Litteram Magistri, Dubio II):

> Item quaeritur de hoc quod dicit: *Unam nativitatem ex Patre veneramur, quae est supra causam, rationem, tempus et naturam.* Sed *contra*: Generatio Filii Dei ex Patre est generatio naturalis; sed generatio naturalis est secundum naturam, non supra naturam: ergo etc.: ergo falsum dicit, quod sit supra naturam. [Also asked about the statement: The birth of the One we worship as the Father, which is above cause, reason, time, and nature. On the contrary, the generation of the Son of God out of the Father is a natural generation; but a natural generation is according to nature, not above nature: therefore, what he says is false, that it is above nature.]
>
> Iuxta hoc quaeritur: penes quae distinguuntur illa quatnor, quae in auctoritate proponuntur. Videtur enim, quod non sit nisi verborum inculcatio. [Accordingly it is asked: By what are those four, which are set forth in the authorization, distinguished? For it seems, that it is nothing but a flourish of words.]
>
> Respondeo: Dicendum, quod natura non accipitur ita communiter, sed pro natura creata; unde non vult dicere, quod generatio Filii sit supra naturam aeternam, quae est natura naturans, sed super naturam creatam, que consuevit dici natura naturata[4]. [Response: It is answered, that nature [here] is not taken generally, but instead as created nature (*natura creata*); so it cannot be said that the nature of the generation of the Son is above the eternal nature, which is the nature naturing (*natura naturans*) [the nature giving birth], but above created nature (*naturam creatam*), the ordinary nature natured (*natura naturata*), the nature born].

I thank Br. Alexis Bugnolo for the translation. See also *Opera Omnia*, index, vol. 10, p. 220.

38 Bonaventure, *Commentaries on the Four Books of Sentences of Master Peter Lombard, Opera Omnia*, vol. 3, p. 197 [p. 218 online]. Distinction 8, Article II, Question III, Conclusion, Doubts on the Writings of the Teachers, Doubt II.

> Response: It is answered, that nature is not taken generally, but instead as created nature (*natura creata*); so it cannot be said that the nature of the generation of the Son is above the eternal nature, which is the nature naturing (*natura naturans*), but above the nature created (*naturam creatam*), the ordinary nature natured (*natura naturata*).

I thank Br. Alexis Bugnolo for assistance with this translation.

39 Bonaventure, *Opera Omnia*, ibid., vol. 3, p. 197, note 4: *Haec naturae distinctio fundamentum est totius libri condemnati Erigenae de Divisione naturae (cfr. II, Sent. d. 1. p. 1. a. 3. q. 2. in scholio). Eadem tangitur etiam ab August., XV. de Trin. c. 1. n. 1: "Deus est natura, scilicet non creata, sed creatrix". Cfr. ibi XIV. c. 9. n. 12. et II. de Anima et eius origine, c. 3. n. 5.*

62 *Autonomous Nature*

This is the fundamental distinction concerning nature that is the basis of the condemned book of Erigena (cf. *Sent*, Bk. II. d. 1. p. 1. a. 3. q. 2, in the scholium). It is also touched upon by Augustine, *On the Trinity* [*De Trinitate*], Bk XV, ch. 1, n. 1: 'God is Nature, not created (nature), but (Nature) the Creatrix' ['*Deus est natura, scilicet non creata, sed creatrix*'], cf. Bk. XIV, ch. 9. n. 12, and *On the Soul and its Origin* [*de Anima et eius origine*], Bk. II, ch. 3, n. 5.

I thank Br. Alexis Bugnolo for the translation.

40 Bonaventure, "Breviloquium," in *The Works of Bonaventure*, trans. Jose de Vinck, 4 vols. (Paterson, NJ: St. Anthony Press, 1963), vol. 2, quotations on p. 69.

41 Bonaventure, "Breviloquium," ibid., pp. 38–40.

42 Bonaventure, "Breviloquium," ibid., pp. 35–37, quotations on pp. 35, 37.

43 Bonaventure, "Breviloquium," ibid., pp. 82–83, quotation on p. 82.

44 Bonaventure, "Breviloquium," ibid., pp. 76–77, quotation on p. 77.

45 Bonaventure, "Breviloquium," ibid., pp. 109–111, quotations on p. 110.

46 Bonaventure, "Breviloquium," ibid., pp. 112–114, quotations on pp. 112 and 113.

47 Lucks, "*Natura Naturans—Natura Naturata*," op. cit., on Beauvais, Lull, and Abano, see pp. 3–4; on Sannig, John of St. Thomas, and Francis Bacon, see pp. 6–8; on Bruno and Spinoza, see pp. 10–12.

48 Rohr, "Man and Natural Disaster," p. 138; Konrad von Megenberg, *Buch der Natur*, ed. Franz Pfeiffer (Stuttgart: K. Aue, 1861).

3 Nature Personified

Renaissance Ideas of Nature

In 1348, a "most terrible plague" hit the city of Florence, Italy. In his *Decameron* (ca. 1351), Italian writer Giovanni Boccaccio (1313–1375) described the disease "as sent from God as a just punishment for our sins." Death, destruction, and the breakdown of society ensued. Neither cause nor cure was known. Nothing seemed to stop its advance. Boccaccio wrote:

> No doctor's advice, no medicine could overcome or alleviate this disease. An enormous number of ignorant men and women set up as doctors in addition to those who were trained. Either the disease was such that no treatment was possible or the doctors were so ignorant that they did not know what caused it, and consequently could not administer the proper remedy. In any case very few recovered; most people died within about three days of the appearance of the tumours . . . most of them without any fever or other symptoms.

The onset of the disease was swift, its course painful and debilitating, its spread rapid and wide.

> The violence of this disease was such that the sick communicated it to the healthy who came near them, just as a fire catches anything dry or oily near it. And it even went further. To speak to or go near the sick brought infection and a common death to the living; and moreover, to touch the clothes or anything else the sick had touched or worn gave the disease to the person touching. . . .

Not only was the physical pain of the disease nearly intolerable, the fear and psychological trauma were almost equally unbearable.

> One citizen avoided another, hardly any neighbour troubled about others, relatives never or hardly ever visited each other. Moreover, such terror was struck into the hearts of men and women by this calamity, that brother abandoned brother, and the uncle his nephew, and the sister her brother, and very often the wife her husband. What is even worse and nearly incredible is that fathers and mothers refused to see and tend their children, as if they had not been theirs.[1]

Figure 3.1 Black Death, 1348.
Illustration of Bubonic Plague in a Bible from 1411.
Source: German *Wikipedia*.

The Black Death, or Bubonic Plague, which ravaged Europe in waves from 1348 through the mid-fifteenth century, brought terror to populations everywhere. Outbreaks occurred in 1360–1361, 1369, and 1374. Appearing at a time when food shortages and famine had already weakened the populace, it attacked without regard to wealth or social status. Later understood to be transmitted by rats and fleas, it devastated cities and countryside alike. Its symptoms were the appearance of black bulblike swellings on the body leading to a painful death within about a week. Other variants attacked the lungs and blood systems.[2]

In this chapter I examine the economic, technological, and intellectual changes that took place from the late Middle Ages to the Renaissance. I show how they interacted with natural disasters and with the feminization and unpredictability of nature. I argue that Greco-Roman conceptions of gender, along with Christian ideas, played a significant role in Renaissance perceptions of nature as female, unruly, and recalcitrant. This framework forms the background to seventeenth-century concepts of experimentation and the laws of nature as a means of understanding and controlling active nature that will be considered in subsequent chapters.

Medieval Economies

By the time the great plague hit Europe in 1348, a series of famines had already decimated lives and livelihoods across Europe. These disasters began with the Great

Famine of 1315–1317, occurring at a time when the warm period of the Middle Ages was drawing to a close and the Little Ice Age—the period from around 1350 to 1850—was beginning. In 1303, 1306, and 1307, the Baltic Sea froze, glaciers expanded, and northern grain production was curtailed. In 1314 heavy rains initiated a series of cold, wet winters. In 1318 an unknown disease (possibly anthrax) attacked sheep and cattle reducing meat supplies and increasing malnutrition. The famines continued during the ensuing decades. Starvation and disease weakened the labor force, reducing harvests and diminishing food stores. Populations all over Europe were destroyed, some declining by as much as 60 percent with the lowest levels occurring during the first half of the fifteenth century.[3]

During the one hundred years between 1350 and 1450, while population levels were at their lowest ebb, marginal lands were abandoned, forests grew back, and soils and pastures recovered fertility, restoring much of the pre-plague ecology in both the European north and south by the early to mid-sixteenth century. But as European populations recovered, the preindustrial economy began to grow again. From around 80 million in 1300, population expanded to around 200 million by the late 1700s.

As these changes occurred, forests again became fields and expanding towns increased pressure to create arable lands. Forests were cut for lumber, wood was dried and smelted in kilns under high heat to make charcoal for fuel, while lands were cleared for pasture and croplands. The European tetrad of field crops—wheat, oats, rye, and barley—which could be sown by broadcast, rather than planted by hand, along with the invention of the heavy plow and horse collar, allowed the wet soils of northern Europe to be cultivated, harvested, and stored as surpluses for food and fodder. The three-field system, in which two fields were planted with grains, while one was left fallow to recover its nutrients, enabled farms to support families and livestock. Domesticated cattle, horses, pigs, goats, and sheep (called the Major Five by Jared Diamond) provided protein and animal labor. Wind and water were harnessed for energy and canals, and dams were built to increase transport.[4]

An economy based on market-trading that arose in the city states of Italy made its way northward accentuating rural-to-urban and city-to-city exchanges of goods and services, while the spread of money provided a uniform medium of exchange and fostered open-ended accumulation. Shipping routes among European nations and explorations of more distant lands worked to facilitate long-distance trade. Cities burgeoned as trading and handicraft centers were stimulated by a growing class of entrepreneurs who contributed to the emergence of nation states.

From the fourteenth through the fifteenth centuries, the economy of Europe was transformed from a medieval manorial economy to a preindustrial economy. Over the course of some two hundred years, material conditions were created that increasingly allowed for the possibility of controlling nature. These developments included the production of agricultural surpluses, technologies for managing wind and water, fortified cities, and long-distance trade—conditions that began to create a buffer for humanity against unruly nature.

But economic advances and the exploitation of human and natural resources interacted dialectically with a nature that could not be wholly harnessed or

66 *Autonomous Nature*

controlled. Despite early economic and technological developments that allowed humanity to achieve increasing management of its food supplies, the unpredictable vicissitudes of nature caused major disruptions to daily life. Indeed the impacts of natural disasters were made all the more devastating by the very successes of human actions and technologies that produced larger surpluses, expanding populations, and ecological disruptions to forests, fields, and waterways. While some disasters, such as earthquakes and volcanoes were fairly localized, others such as plagues and famines could be widespread. Moreover, unpredictable weather patterns such as droughts, freezes, and storms limited food storage and weakened humanity's shields against nature.

Preindustrial Capitalism

During the sixteenth century, under preindustrial capitalism, further buffers against the ravages of unruly nature were created through new technologies, early industries, and trade routes. The mining and trading of metals expanded, commercial banking organizations acquired mines and foundries, and new metallurgical industries created products for agriculture, commerce, and warfare. Iron smelted in blast furnaces and hammered in forges was used to make farming implements such as plows, bolts, tools, and horseshoes; household utensils such as pots and pans; and military impedimenta such as cannons, guns, and ballistics. With the depletion of forests, coal mining grew exponentially, leading to new ecological and human disruptions, such as air and water pollution, mining accidents, and illnesses.

Preindustrial capitalism created class divisions that benefited landlords and well-to-do farmers, but also contributed to a new class of wandering "masterless men" and squatters who lost lands through enclosures and the inability to pay higher rents. Some of the landless became laborers in the new industries of shipbuilding, mining, and metal refining or sought employment as apprentices in growing cities.

An agricultural improvement movement capitalized on the changes in the countryside. In the northern states, farmers who introduced new agricultural techniques, such as fertilizers, legumes, crop specialization, and new field rotation systems, reaped profits from surpluses exchanged on the market. In England, landlords enclosed land from the commons to increase holdings, cut forests for pasture and cultivation, and drained fens to reclaim land. Yeoman farmers produced wool for textiles and crops for the urban trade. In the Netherlands, liquid manure, peat ash, and clover contributed to higher yields of wheat, rye, beans, and dairy products. Vegetable crops (such as onions, carrots, turnips, and cabbages) and fruits (such as cherries and strawberries) diversified production systems and diets. Such innovations helped to intensify production and increase yields while also creating buffers that offset the devastations caused by famines, droughts, and storms.

Despite such achievements, however, unpredictable events and natural disasters disrupted daily life and social institutions. Diseases, such as the plague, smallpox, measles, scarlet fever, and pneumonia, for which no cause or cure could be found, were often attributed to the wrath of God acting through nature in retaliation for human sins. If humans failed to act in accordance with God's law, "He" could render reprisals in the form of crop failures, droughts, storms, and earthquakes. An earthquake in Ferrara, Italy, in 1570, for example, elicited assertions that earthquakes were the voice of God and that it was "one of God's ways of admonishing the wicked and an exhortation to repentance." Biblical passages cited included Psalms 13:3, which stated, "At his wrath the earth trembles," and Leviticus 26 which warned, "If in spite of these things, you will not be corrected but will walk contrarily with me, then I will smite you, even I, seven times for your sins."[5] But other interpreters, drawing on ancient Greco-Roman myths and philosophies, saw such disasters as the direct result of Nature's rebellious actions.

The Personification of Nature

In direct opposition to economic advances, and in seeming defiance of such achievements, was an apparently rebellious Nature, personified as a woman who manifested "herself" in natural disasters—famines, droughts, storms, volcanic eruptions, earthquakes, and epidemics—wreaking havoc and chaos on rich and poor alike. Harkening back to Hesiod's *Theogony*, Plato's *Timaeus*, and Aristotle's *Physics*, the feminization of nature and matter and the masculinization of intellect and form set the stage for explaining disruptions and destructions (see Chapter 1). By the era of the Renaissance, female Nature, which had a long-standing association with chaotic unpredictability and female recalcitrance, stood as a scapegoat for both human ills and natural disasters.[6]

During the Renaissance, the goddess *Natura* continued the ancient identity of nature as not only motherly and kind, but as chaotic, unpredictable, recalcitrant, and unruly. Exemplifying the medieval distinction between "nature naturing" and "nature natured," *Natura* became the vice-regent of God, carrying out God's will in the natural and human worlds. But although *Natura* reflected the natural order of God in the larger cosmos, "she" sometimes acted in the earthly world on her own, creating unpredictable outcomes, with disastrous results for humanity's health and survival.

Between the twelfth and sixteenth centuries a cosmology developed that personified Nature as the lower form of the World Soul and assigned her the role of carrying out God's dictates in the natural world. Deriving from Neoplatonic sources, *Natura* represented a nature that could be obedient and rational, but sometimes recalcitrant and willful.

The tensions between the orderly and unruly personalities of Nature appeared in late medieval and Renaissance philosophy, art, and literature. These aspects of nature were present in the cosmology of Bernard Silvestris and Nicholas of Cusa; the art of Leon Battista Alberti, Leonardo da Vinci, and Albrecht Dürer; the politics of Richard Hooker; and the philosophy of Giordano Bruno. Together they illustrate the ways in which nature was gendered and personified as both orderly and rebellious.

68 *Autonomous Nature*

Bernard Silvestris

Bernard Silvestris (1085–1178) was the author of the influential *Cosmosgraphia* (*De Mundi Universitate*), written ca. 1143–1148 in the twelfth century, a book on the creation of the world narrated from a Platonic and Neoplatonic perspective. The *Timaeus*, the only dialogue of Plato known at the time (in the incomplete Latin of Calcidius), was the inspiration behind the *Cosmographia*. In it, *Natura* and other goddesses are personified and the narrative of creation follows the storyline of Plato and Plotinus, with both the unchanging Ideas and the opposing forces of change depicted primarily as female deities. Divine Providence is represented by *Noys*, Nature by *Natura*; Primoridal matter by *Silva* (Greek, *Hyle*); the World Soul by *Endelechia*—the Bride of *Mundus*, the world; *Urania*, the celestial principle; and *Physis*, the material principle. *Genius*, counterpart of the Demiurge, imprints the celestial forms on matter.[7]

Issues of unpredictability and control are central to the *Cosmographia*. The problem was that disorder and chaos were rampant in the world. It is necessary that Form and order be imposed on matter to control chaos. *Natura's* complaint (title of Alain of Lille's subsequent allegory, *De Planctu Naturae*, ca. 1160) was answered by *Noys's* creation of the four elements out of primordial matter and the creation of the World Soul, *Endelechia*. The sequence is reminiscent of Plotinus's emanation (or rebellious audacity) in the separation of Nous from Intellect followed by the World Soul and subsequent descent of the individual souls.

After the creation of the heavens and lower world—the Megacosmus (Macrocosm)—comes the creation of "man"—the Microcosmus (Microcosm). *Natura* creates humanity out of *Urania*, the celestial principle, and *Physis*, here characterized as the material principle (which like *Natura* and *Urania* is female), with the help of Genius, who impresses the forms on matter. *Urania* and *Natura* descend to earth bringing with them *Endelechia*, the World Soul, where they encounter *Physis*. *Natura* then unites *Urania*, who supplies portions of the World Soul to create human souls, with *Physis*, who supplies the body. Together the three female personalities, *Urania*, *Natura*, and *Physis*, jointly create "man." Because humans (and other organisms) can create new life, the universe will continue its own regeneration (creative activity) and not lapse back into chaos. Order (predictability) is thus imposed on chaos (unpredictability).

In discussing the significance of Bernard Silvestris's innovative account, Peter Dronke writes:

> However expressively *natura* may feature in medieval Latin writings before Bernard Silvestris, there is no question of a fully-fledged 'goddess Natura'. . . . Nowhere before Bernard, for instance, is Natura seen as 'the blessed fecundity of the womb' of the goddess Noys; nowhere previously had Natura been distinguished from Physis, whom Bernard presents as a more modest cosmic artisan. Nowhere before Bernard is Urania, who was known as the Muse of astronomy, transformed into a higher cosmic *creatrix*—Natura's empyrean double. Nor can I trace before Bernard his triad of nature-goddesses (Urania,

Natura, Physis), creating man. . . . The whole conception of a feminine *trinitas creatrix*, a counterpart to the celestial Christian trinity, working at the fabrication of man in a grassy Elysium, is unparalleled.[8]

The personification of *Natura* would set the tone for Renaissance depictions of nature and for ways in which *Natura naturans* and *Natura naturata* would frame the tensions between nature creating and nature as created world.

Nicholas of Cusa

During the fifteenth century, the German Cardinal and philosopher, Nicholas of Cusa (1401–1464), developed a Christian Neoplatonism that encompassed science, mathematics, law, religion, art, and technology. Cusa was influenced by Plato, Aristotle, and Aquinas, among others. In particular, he urged his followers to read Erigena and he used Erigena's language concerning an active, creating God and the created world. Like Erigena, Cusa would later be seen as a pantheist for his view that the created world is one with the creating principle and that God is in everything. And like Erigena his method was rooted in a dialectic of opposites.[9]

Cusa's 1440 work, *On Learned Ignorance* (*De Docta Ignorantia*), set out the interconnected relationships between God as creator, the created world, the human being, and Christ. "Learned Ignorance" is a manifestation of the dialectic between knowing and unknowing, being and becoming, infinite and finite. As a "coincidence of opposites," God is an infinite Oneness, comprising both maximum and minimum. The created world is enfolded within the original Oneness and unfolds from it. Each individual in the created world flows dialectically outward from the One and is encompassed within it in a continuing dynamic process. The One is a whole, comprising a totality of interrelated, interconnected parts. Each thing is a part of everything and everything in turn comprises each thing. All things are in constant motion and the universe is constantly changing. There is no center to the universe and no boundaries. The center and circumference are in God and God is everywhere and nowhere.[10]

Regarding humanity, Cusa held that the human being is within God and inseparable from him. Christ is both human and divine. The Divine Mind's knowing process creates beings. Things in the created world are thus material images of abstract truths in the Divine Mind. The human mind as an image of the Divine Mind creates concepts. Truths in the Divine Mind are mirrored in the human mind, mathematical concepts being truths that are not subject to diminution or change.

Hans Blumenberg points out that the craftsman, featured in Cusa's dialogue, *Idiota de mente* (*The Layman: About Mind*) of 1450, holds the key to a new idea of human ingenuity stemming from the mind's creative power beyond the mere imitation (*imitatio*) of nature. The illiterate maker of spoons, plates, bowls, and jugs does so, not from imitating nature, but as a creative idea that stems from imitation of the mind of God. Rather than completing what nature would have done if "she" could grow beds and tables from wood, "Man," as made in God's image, creates

70 *Autonomous Nature*

new things in the created world by imitating ideas in the mind of God. Here, in the distinction between the *imitatio* of God and the *imitatio* of nature, emerges "the antagonism between the mechanical and the organic, art and nature. . . ." Erigena's differentiation between *natura creans* (*naturans*) and *natura creata* (*naturata*) is exemplified as creating through the mind and the resultant created objects in the sensible world.[11]

The distinction between creating artist and created objects in the Renaissance era rests on the foundations of both Christianity and Neoplatonism, God and Ideas, as well as on Aristotelianism and the imitation of nature. These foundations would be borne out in the art and artisanry of the great painters, poets, and technicians who led the way forward to the Scientific Revolution.

Natura in Art and Architecture

The imitation of nature, *Natura*, by the creating artist was at the root of Renaissance art in the fifteenth century. Leon Battista Alberti (1404–1472), sculptor, painter, and architect wrote that "all levels of learning must be claimed from nature herself." Sculptors and painters alike should "take all things from Nature herself" and "always imitate Nature." "Let us always take from Nature [objects] that we wish to paint and from them always let us select the most beautiful and the most deserving."[12] In Alberti's work, imitation embraced both meanings of nature as *natura naturans*, or nature as active and creating, and *natura naturata*, or nature as created world. Nature appeared in the objects of the natural world (the passive results) that the artist rendered on canvas and as the creative power behind the visible world, nature's active first principles, that inspired the artist. As art historian Jan Bialostocki put it: "[T]he simple, natural appearance of real things in the world that surrounds the artist—*il naturale*—is one aspect of nature. The other is formed by the concept of a mighty power, of *natura naturans*."[13]

For Alberti, Nature as active creator delighted in painting. But the artist, in following nature's first principles, might be able to achieve even greater perfection than nature "herself," inasmuch as the visible world was always filled with imperfections. According to Bialostocki,

> Both imitation of *natura naturata* and of the creational powers of nature have been known to previous times, but in supposing that an artist who creates a work of art is able to achieve a greater perfection than nature in her work has ever achieved, Alberti formulated a new idea of great importance.[14]

While medieval writers had considered nature to be superior to human artists, Alberti believed it possible for human artists to surpass nature's artistry. Art historian Erwin Panofsky, in interpreting this relationship of art to nature, wrote that "classical art itself, in manifesting what *natura naturans* had intended but *natura naturata* had failed to perform, represented the highest and 'truest' form of naturalism."[15]

Mary Garrard has analyzed the creating and created dimensions of fifteenth-century art in terms of the female figure, *Natura*, in the paintings of Leonardo da Vinci (1452–1519). Her perspective stems from the idea that the medieval distinction that remained in force throughout the [century] was that between *natura naturans* and *natura naturata*, between the dynamic, generative principle and the inert material result of that creation. . . . In this way of thinking, art was not to imitate the mere phenomena of nature (*natura naturata*) but, rather, its higher invisible principles (*natura naturans*); and by giving them perfected visible form, art could successfully compete with Nature herself.[16]

Here nature is active and creative, inspiring the artist to imitate "her" perfection through his art. In keeping with the influence of Neoplatonism, this perspective was in direct opposition to the Aristotelian distinction between active male form and passive female matter. But for Leonardo, "the creative and material [aspects of] nature were . . . conflated into a single entity, a female Nature" with "creative powers."[17] In Leonardo's paintings, artistic creation stems from the artist's knowledge of and unity with nature's inherent law. "Creating" nature, however, could be both positive—reflecting nature's generative powers—and negative—reflecting its violence and destructive potential.[18]

In the *Virgin of the Rocks* (Louvre, from the 1480s), "creating" nature (*natura creans*) appears within matter rather than being transcendent above it. Living, growing, animate nature is depicted in the swirling growth patterns of the plants at the base of the painting. The virgin herself links female nature with its generative capabilities to the timelessness of the virgin's own potential for generation captured and preserved by the perfection of the artist. Similarly, in the portrait of Cecilia Gallerani (ca. 1484–1485), a beautiful woman in three-quarters pose holds an ermine, symbol of the Dukedom. Here again the pure forms of mathematics are integrated with the living world of human and animal. As Garrard states, her "prominent hand exhibits the force of ideal structure, rhyming in shape with the animal under its control, and thereby suggesting that the sharp, fierce creature might stand for an aspect of her personality. . . ."[19]

The *Mona Lisa* illustrates the timeless perfection of an arresting woman who will live forever as a portrait, but who is also integrated into the backdrop of changing nature, worn down by the violent forces of water. Linking the portrait and its background reinforces the analogy between woman and nature. "The woman's face and the landscape background together express the processes of nature as symbolized by the image of a female, whose connection with human generation links her sex with the creative and destructive powers of nature." Nature, like woman, had anatomical, circulatory, and generative systems of birth, growth, death, and decay. "The earth has a spirit of growth," said Leonardo, "whose flesh is the soil, whose bones are mountains, whose blood is its waters."[20]

Leonardo's anatomical drawings of female reproductive organs (the "Great Lady," ca. 1510) accompanied by his annotations place the active generative power of human reproduction in the female womb. The power of the female thus challenged prevailing Aristotelian beliefs that the female contributed only passive matter on which the active male semen worked to create new life. Not only did the female womb have creating, generative power, but nature's womb could overpower humanity's creations.[21]

72   *Autonomous Nature*

Several of Leonardo's paintings celebrated nature's rebellious, willful, uncontrollable side. In his drawings of the "Deluge" (ca. 1517–1518), storms and flooding (*natura naturans*) overpower weaker humanly constructed houses and other artifacts (*natura naturata*), once again recognizing the generative power of active nature. As Garrard puts it:

> In those images of ferociously spiraling explosions of water, a giant apocalyptic flood repeatedly destroys the tiny structures of cities and towns. It is a celebration, rare in Italian Renaissance art, of the superior power of nature over human civilization—implicitly also a gender construct—of the endurance of female generation over male culture.[22]

Albrecht Dürer (1471–1528) symbolizes a transition to a more objective stance toward nature. In the "Artist Drawing a Nude with Perspective Device" from his *Painter's Manual* (1525), the artist views a reclining female figure, symbolic of nature as woman, through a grid that allows for a focused, but distant perspective. Here the artist is not only separated from female nature by the frame, but the reflection of the frame's grid on the artist's table puts him in a position of superiority over the object.

Here perspective art reflects the mathematical certainty and truths of Platonic and Neoplatonic philosophy as dominant over passive, material nature. Dürer's linear perspective creates both the naturalism and realism that links art and science in ways that would be more fully articulated in the geometrical and mathematical approaches of the seventeenth-century Scientific Revolution.[23]

Natura in Politics

Richard Hooker's (1554–1600) major work, *Of the Laws of Ecclesiastical Polity* (1593), drew on Nature as God's representative on earth and God as the source of natural law. Nature stemming directly from God was the source of human knowledge. God as creator spoke and taught through created nature on earth. Nature, however, was God's instrument and conveyor; she did not have power or a will of her own. "For that which all men have at all times learned, Nature herself must needs have taught; and God being the author of nature, her voice is but His instrument." Created nature was a reflection God's message to humanity.[24]

For Hooker, God's reason gave rise to natural law, which in turn was the foundation of earthly societies and government. His emphasis on natural law stemmed from his reading of Augustine and Aquinas. God's law operated rationally both in nature and in society. But although natural law was unchanging and immutable, society's laws could be changed according to circumstances and need. Law's "seat is the bosom of God, her voice the harmony of the world, all things on earth do her homage . . . admiring her as the mother of their peace and joy." His theory that government was rooted in natural law and applied both to created nature and to human society would later become a basis for John Locke's *Two Treatises on Government*.[25]

Renaissance Ideas of Nature 73

By insisting that Nature was God's instrument, Hooker maintained an ortho-doxy concerning Nature's potential for rebellious action. For him, the World Soul of Plato and Plotinus did not have the potential for independent action. As Eustace Tillyard put it,

> Hooker, orthodox as usual, is explicit on this matter. She [Nature] cannot be allowed a will of her own or the rank of a kind of goddess. She is not even an agent with her eye ever fixed on God's principles; rather she is the direct and involuntary tool of God himself.[26]

Hooker thus stood against a trend of his time, stemming from Marsillo Fici-no's revival of Neoplatonism, that assigned a soul to nature. Plotinus had indeed described the separation of the World Soul from the original Oneness in terms of its own rebellious audacity (*Tolma*). Concerning the Neoplatonic tendencies of the time, Tillyard notes:

> This giving a soul to nature—nature, that is, in the sense of *natura naturans*, the creative force, not of *natura naturata*, the natural creation—was a mildly unorthodox addition to the [levels of] spiritual or intellectual beings. ... For the Elizabethans talked much about nature, and she cannot be omitted from the world picture. That there was a law of nature was universally agreed; she worked unswervingly by a set of rules applicable to her alone; but the question still remained whether she was a voluntary or involuntary agent.[27]

It was through the voluntary action of Nature that Renaissance thought accounted for unexplained, seemingly irrational events in the created world. Crop failures, droughts, earthquakes, and the like were both manifestations of the will of nature to act independently of God and the failure of humanity to follow God's rational path. Here Will has the upper hand over Reason. The tension between Reason and Will, or rationality and voluntarism, was at the root of much of the ensuing theological and scientific controversy over the nature of God that took place in the seventeenth century.

Giordano Bruno

The work of Giordano Bruno (1548–1600) marks the transition to the Scientific Revolution of the seventeenth century and the beginnings of Bruno's influence on major figures such as Baruch Spinoza and Gottfried Wilhelm Leibniz. Like Erigena and Cusa before him and Spinoza after him, Bruno was identified with what became known as pantheism. Bruno synthesized his philosophy from many sources, including the Presocratics, Neoplatonists, and Hermeticists, along with Aristotelianism and Christianity. Like Erigena, he held that God (the Universe) was the active creator or force, *natura naturans* (*creans*), while the created or phenomenal World was *natura naturata* (*creata*). In his cosmology, like Nicholas of Cusa and Copernicus, he argued for a sun-centered, infinite universe and a

74   Autonomous Nature

plurality of worlds in which stars were surrounded by planets. The universe was not hierarchical and there was no special heaven or place of God. Instead, God was immanent in everything and everywhere present.[28] Although Spinoza knew Bruno's work through his teacher Frans van den Enden, it is not clear from what sources other than the "Thomists," Spinoza obtained the terms *natura naturans* and *naturata* (see Chapter 5).[29]

In his 1584 book, *Cause, Principle, and Unity*, Bruno used the term *natura naturans*. Alfred Weber writes in his 1896 *History of Philosophy* that in order to escape the charge of atheism, Bruno distinguished between the Universe (God) and the World (phenomena). God was the universe or *natura naturans*, the eternal, infinite cause of the world. The phenomenal world itself was the totality of God's effects, the *natura naturata*. The Universe had no beginning or end, whereas the phenomenal World lived and died.[30]

Nevertheless, for Bruno, God was actually the Soul of the World. God was its immanent, rather than transcendent, cause—a position that ultimately led to the charge of atheism, conviction, and death. Weber argues that

> the God of Bruno is neither the creator nor even the first mover, but the soul of the world; he is not the transcendent and temporary cause, but, as Spinoza would say, the immanent cause, i.e., the inner and permanent cause of things; he is both the material and formal principle which produces, organizes, and governs them *from within outwardly*; in a word, their eternal substance. The beings which Bruno distinguishes by the words, "universe" and "world," *natura naturans* and *natura naturata*, really constitute but one and the same thing. . . . [31]

Bruno drew on Aristotle's distinction between form as male and matter as female as well as Biblical associations of eternal life and the Fall from Eden to express the process of unification as a sexual conjugation. Form was the unchanging constant truth; matter the changing, inconstant substrate. Form was associated with Adam, Paradise, and the Tree of Life; matter with Eve, unpredictability, and the loss of Paradise. Significantly, it is through sexual union that form and matter are conjoined. Substance itself is actually a pantheistic unity of *natura naturans* and *natura naturata*. The two are intimately coupled together in eternity. Bruno wrote:

> That is why form, symbolized by the man, entering into intimate contact with matter, being composed or coupling with it, responds to the *natura naturans* with these words, or rather this sentence: '*Mulier, quam dedisti mihi*', *idest*, matter which was givenme as consort, *ipse me decepit; hoc est*, she is the cause of all my sins. Behold, behold, divine spirit, how the great practitioners of philosophy and the acute anatomists of nature's entrails, in order to show us nature plainly, have found no more appropriate way than to confront us with this analogy, which shows that matter is to the order of natural things what the female sex is to economical, political and civil order.[32]

Renaissance Ideas of Nature 75

For Bruno, Nature (*Natura naturans*) is self-active. God/Nature, Male/Female, Being/Process, Identity/Change, Order/Disorder are contained within a single active, changing substance, or Soul of the World. Matter (change) is an intimate part of the order of nature, just as change (symbolized by the female) is an intimate part of economics, politics, and civil society. Fundamentally, both the natural and social orders are constituted by change and process. Natural change and political change are both primary aspects of the real world. These ideas were revolutionary in the context of his times. They would lead to fatal consequences for their author.

For Bruno, Nature is the Creator which both necessarily and voluntarily produces all things. Power and Will, Freedom and Necessity, Matter and Form, are all one and the same, operating as immanent, productive forces. The Universe unfolds into the World. God as Infinite Being is everywhere and in everything. In his 1591 book, *On the Monad, Number and Figure* (*De Monade, numero et figura*), Bruno described three types of monads: God, souls, and atoms. Here the Universe/God/the One/the Monad is the living Soul of the World and this World Soul is everywhere alive.[33]

Everything in nature is alive. God is immanent in every rock, blade of grass, and every earthly being. A seed becomes a plant, the plant feeds an animal, the animal feeds a human who lives, dies, and returns to the earth. The One, or Monad, expands and contracts, becomes and dies, is reborn and dies again.[34]

In 1592, Bruno was imprisoned in Rome, charged with holding opinions contrary to the Catholic faith, and tried. The charges included beliefs against the Trinity, the virginity of Mary, the divinity and incarnation of Christ, and belief in magic. He was burned at the stake in 1600.[35]

Conclusion

Between the late Middle Ages and the end of the sixteenth century, a preindustrial economy emerged that began to give humanity the means of controlling nature through crop surpluses, agricultural fertilizers, energy from coal and charcoal, and new metallurgical implements. These economic and technological advances helped to offset unpredictable natural events that periodically intervened to threaten human lives and livelihoods.

The intellectual framework that developed during the Christian era, depicting nature as having a creating and created aspect (identified as *natura naturans* and *natura naturata*), was elaborated and exemplified during the Renaissance through cosmology, art, architecture, politics, and philosophy. Nature was enlivened as the goddess *Natura*, the lower form of the cosmic World Soul. As such, "she" conveyed God's will to the phenomenal world. God's pleasure or displeasure with humanity was displayed in crop harvests or failures, pleasant or harsh weather, and human health or disease. But Nature could also act, *not* as the instrument of God, but on her own as a willful and rebellious female who disrupted human affairs and created natural disasters.

These depictions of nature as alive and unruly raised the issue of control and management. Through mathematics and experimentation, developed during the

76 *Autonomous Nature*

seventeenth-century Scientific Revolution, nature's actions could be understood and predicted. With prediction came the possibility of human control over natural phenomena. The idea of nature as a machine made up of dead, inert parts that could be repaired and replaced superseded the concept of a living, unruly, female nature. The work of Francis Bacon and Robert Boyle, among others, led to natural knowledge gained through experiments, while the systems of Descartes, Spinoza, Leibniz, and Newton led to a new understanding of nature's mathematical regularities. These two approaches, elaborated in the next three chapters, allowed for the prediction and control of nature.

Notes

1   Giovanni Boccaccio, *The Decameron*, trans. Richard Aldington (Garden City, NY: International Collectors Library, 1930); Robert S. Gottfried, *The Black Death* (London: Robert Hale, 1983).

2   Carlo M. Cipolla, *Before the Industrial Revolution: European Society and Economy, 1000–1700* (New York: W.W. Norton, 1976), Ch. 5, pp. 146–157.

3   Barbara W. Tuchman, *A Distant Mirror: The Calamitous Fourteenth Century* (New York: Knopf, 1978), pp. 24–25; Cipolla, *Before the Industrial Revolution*, pp. 199–201.

4   The following sections draw on Cipolla, *Before the Industrial Revolution*, Ch. 6, pp. 158–181; Bruce M.S. Campbell, ed. *Before the Black Death: Studies in the "Crisis" of the Early Fourteenth Century* (Manchester, UK: Manchester University Press, 1991), Lynn White, Jr., *Medieval Technology and Social Change* (London: Clarendon Press, 1962); Carolyn Merchant, *The Death of Nature: Women, Ecology, and the Scientific Revolution* (San Francisco: HarperCollins, 1980), pp. 42–68 and *Reinventing Eden: The Fate of Nature in Western Culture*, 2nd ed. (New York: Routledge, 2013), pp. 58–59.

5   Joanna Weinberg, "'The Voice of God': Jewish and Christian Responses to the Ferrara Earthquake of November 1570," *Italian Studies* 46 (1991): 69–81, esp. pp. 70, 78–80, quotations on pp. 78, 79; Craig Martin, *Renaissance Meterology: Pomponassi to Descartes* (Baltimore: Johns Hopkins University Press, 2011); Amelia Kikue Linsky, *The Ferrara Earthquakes, 1570–1579: Science, Religion, and Politics in Late Renaissance Italy* (Middlebury, VT: Middlebury College, April 22, 2013).

6   Merchant, *Death of Nature*, pp. 127–143.

7   Bernard Silvestris, *Cosmographia*, ed. Peter Dronke (Leiden: E.J. Brill, 1978), pp. 65–66.

8   Peter Dronke, "Bernard Silvestris, Natura, and Personification," *Journal of the Warburg and Courtauld Institutes* 43 (1980): 16–31, quotation on pp. 19–20.

9   Jasper Hopkins, *A Concise Introduction to the Philosophy of Nicholas of Cusa* (Minneapolis, MN: A.J. Banning Press, 1986), pp. 5–6, 37, 42.

10 Nicholas of Cusa, *On Learned Ignorance* (*De Docta Ignorantia*), trans. Jasper Hopkins (Minneapolis, MN: A.J. Benning, 1981); Cusa, *Selections, English and Latin* (Cambridge, MA: Harvard University Press, 2008); Hans Blumenberg, "Toward a Prehistory of the Idea of the Creative Being," trans. Anna Wertz, *Que Parle*, special issue on *The End of Nature* 12, no. 1 (Spring/Summer 2000): 17–54, see pp. 19–21, 23, 42–43; Clyde Lee Miller, "Cusanus, Nicolaus [Nicolas of Cusa]," The Stanford Encyclopedia of Philosophy (Fall 2009 Edition), Edward N. Zalta, ed., http://plato.stanford.edu/archives/fall2009/entries/cusanus/.

11 Blumenberg, "Toward a Prehistory of the Idea of the Creative Being," pp. 17–20, 23, 42–43, quotation on p. 23; Nicholas of Cusa, *Idiota de mente* (*The Layman, About Mind*), trans. with an Introduction by Clyde Lee Miller (New York: Abaris Books, 1979).

12 Leon Battista Alberti, *On Painting: A New Translation and Critical Edition*, ed. and trans. Rocco Sinisgalli (New York: Cambridge University Press, 2011), quotations from

Bk. 3, secs. 55 and 56, pp. 77, 79; Alberti, *On Painting and On Sculpture: The Latin Texts of* De Pictura *and* De Statua, ed. with trans., introduction, and notes by Cecil Grayson (London: Phaidon, 1972), see pp. 64–67, 80–81, 86–89, 100–103; Alberti, *On Painting*, ed. John Richard Spencer (New Haven, CT: Yale University Press, 1956), pp. 67, 70, 78; Alberti, *On the Art of Building in Ten Books*, trans. N. Leach, J. Rykwert, and R. Tavenor (Cambridge: MIT Press, 1988).

13 Jan Bialostocki, "The Renaissance Concept of Art and Antiquity," in *The Renaissance and Mannerism: Studies in Western Art: Acts of the Twentieth International Congress of the History of Art* (Princeton, NJ: Princeton University Press, 1963), vol. 2, pp. 19–30, quotation on p. 21.

14 Bialostocki, "Renaissance Concept of Art," quotation on p. 24.

15 Erwin Panofsky, *Renaissance and Renascences in Western Art* (Stockholm: Almqvist and Wiksell, 1960), p. 30, quoted in Bialostocki, "Renaissance Concept of Art," p. 27.

16 Mary D. Garrard, "Leonardo da Vinci: Female Portraits, Female Nature," in *The Expanding Discourse: Feminism and Art History*, ed. Norma Broude and Mary D. Garrard (New York: HarperCollins, 1992), pp. 58–86, quotation on p. 71.

17 Garrard, "Leonardo da Vinci: Female Portraits, Female Nature," p. 72.

18 Bialostocki, "Renaissance Concept of Art," pp. 24–26.

19 Garrard, "Leonardo da Vinci," p. 73, quotation on p. 64.

20 Garrard, "Leonardo da Vinci," ibid., pp. 67–71, quotations on pp. 67, 71.

21 Garrard, "Leonardo da Vinci," ibid., pp. 70.

22 Garrard, "Leonardo da Vinci," ibid., p. 78.

23 Garrard, "Leonardo da Vinci," ibid., p. 72; http://www.oneonta.edu/faculty/farberas/arth/arth200/durer_artistdrawingnude.html.

24 Richard Hooker, *Of the Laws of Eccesiastical Polity* (1593), quotation in *Encyclopedia Britannica*, 11th ed. (London: Cambridge University Press, 1910–11), cf. Hooker, Richard.

25 Hooker, *Laws of Ecclesiastical Polity*, Bk. 1, Ch. 16, "Conclusion"; *Encyclopedia Britannica*, 11th ed., cf. Hooker, Richard.

26 Eustace M.W. Tillyard, *The Elizabethan World Picture* (New York: Vintage, 1959), pp. 45–46, quotation on p. 46.

27 Tillyard, *Elizabethan World Picture*, p. 45, quotation on p. 46.

28 Giordano Bruno, *On the Infinite Universe and Worlds* (*De l'Infinito Universo et Mondi*) (1584). William Turner, "Giordano Bruno," *The Catholic Encyclopedia*, vol. 3 (New York: Robert Appleton Company, 1908), http://www.newadvent.org/cathen/03016a.htm; Giordano Bruno, *Wikipedia*, http://en.wikipedia.org/wiki/Giordano_Bruno.

29 Henry A. Lucks, "*Natura Naturans—Natura Naturata*," *The New Scholasticism* 9, no. 1 (1935): 1–24, see p. 10; John McPeek, *Bruno, Spinoza, the Terms* Natura Naturans *and* Natura Naturata (New Orleans: Tulane University, 1965), p. 35; Dorothea Waley Singer, *Giordano Bruno: His Life and Thought, with Annotated Translation of His Work*, On the Infinite Universe and Worlds (New York: Schuman, 1950).

30 Alfred Weber, *History of Philosophy*, trans. Frank Thilly from the 5th French edition (London: Longmans Green, 1896), "Giordano Bruno," pp. 286–291, see pp. 288–289:

> In order to escape the charge of atheism, Bruno distinguishes between the universe and the world: God, the infinite Being, or the Universe, is the principle or the eternal cause of the world: *natura naturata*; the world is the totality of his effects or phenomena: *natura naturata*. . . . The universe, which contains and produces all things, has neither beginning nor end; the world (that is, the beings which it contains and produces) has a beginning and an end. The conception of nature and of necessary production takes the place of the notion of a creator and free creation. Freedom and necessity are synonymous; being, power, and will constitute in God but one and the same indivisible act.

78   *Autonomous Nature*

31   Weber, *History of Philosophy*, quotation on pp. 288.
32   Giordano Bruno, *Cause, Principle, and Unity and Essays on Magic*, ed. Richard J. Blackwell and Robert de Lucca with an Introduction by Alfonso Ingegno (Cambridge, UK: Cambridge University Press, 1998), Fourth Dialogue, p. 71. See also Giordano Bruno, *Cause, Principle, and Unity*, trans. Jack Lindsay (New York: International Publishers, 1962), p. 118.
33   Giordano Bruno, *De Monade, numero et figura* (Frankfort, 1591) in Bruno, *Opera Latine Conscripta* (Naples: Morano, 1884), vol. I, Part II, 2, pp. 319–484; Weber, *History of Philosophy*, p. 290: "All beings whatsoever are both body and soul: all are living monads, reproducing, in a particular form, the Monad of monads, or the God-universe."
34   Weber, *History of Philosophy*, p. 290.
35   Bruno, *Wikipedia*, http://en.wikipedia.org/wiki/Giordano_Bruno.

Part II
Controlling Nature

4 Vexing Nature

Francis Bacon and the Origins of Experimentation

In 1602, fourteen-year-old Mary Glover fell violently ill after an encounter with Elizabeth Jackson, an older woman who lived nearby. While Mary was visiting Jackson's house, the old woman went into in a towering rage and berated the young girl for meddling with her own daughter's apparel. Three days later, after another encounter with Jackson, this time in her own home, Mary Glover's throat constricted so badly she could not drink. On hearing the news, Jackson delivered the curse, "I hope the Devil will stop her mouth!" For eighteen days, Mary's throat and neck were so badly swollen that she was unable to eat or drink and experienced regular fits of hysteria.[1]

Vexed by the devil from Jackson's curse, Mary Glover experienced violence and torment. Doctors were unable to diagnose or cure her. Neighbors brought Jackson back to Glover's house and forced her to touch the girl in a series of "trials" or "experiments" to prove Mary had been afflicted by Satan. Indeed, Satan, it seemed, caused Glover's tongue to blacken, her face to distort, her head to toss, and her mouth to gape open. She writhed and struggled violently, unable to be restrained. Priests prayed for her deliverance. Prayers and pleadings to God were uttered. Finally, the preacher "urged the Lord to show his power, to give check to Satan, and command him to be gone." But "the more earnest that he was in his prayer, the more she raged in his arms." Finally, on the eighteenth day of the trauma, when taken for dead, her "mouth and eyes opened," she lifted her head, and uttered, "he is come, he is come."[2]

Vexation, defined as possession by the devil, was thought to be induced by witchcraft when outside forces were introduced into the human body. Instances were known throughout the Middle Ages. Violent fits, in which victims had to be restrained by force, marked the trauma and vehemence associated with vexation.[3]

In 1605, three years after the Mary Glover case, Learned Counsel to James I, Francis Bacon (1561–1626), in his *Advancement of Learning*, began using the term "trials and vexations of art." Here the term "trial" meant testing or proving, while "vexation" meant constraining or tormenting and art (*techne*) meant technology. The Glover case in fact contained several significant features that may have influenced Bacon's concept of vexation and possibly its association with physical torment.[4]

Figure 4.1 Demonic Vexation.
A Demon Leaves the Body of a Possessed Woman.
From Pierre Boaistuau, *Histories prodigueses* (Paris, 1597).
In Ernst and Johanna Lehner, *Devils, Demons, and Witchcraft* (New York: Dover, 1971), p. 18, image 39, used by permission.

Bacon used the term "vexation" to mean the constraint and transformation of matter (material bodies) by human means (art). Although Mary Glover's vexation was caused by spiritual intervention (the Devil), her physical body, "writhing and struggling," had to be restrained by human means. Indeed, Mary experienced such trauma and violence during her fits that she had to be held in constraint by human hands. For Bacon, writing his myth of Proteus in 1609, "constraint and binding" was "most expeditiously effected if matter be laid hold on and secured by the hands." Although Bacon did not hold to vexation by devils, he may well have

Vexing Nature 83

Figure 4.2 Francis Bacon (1561–1626).
From *Old England's Worthies*, London: C. Cox, 1847, p. 97 (illustration), pp. 108–110 (biography). Courtesy of Getty Images, Hulton Archive 463984775.

been influenced by the term "vexation" as an inspiration for his concept of experimentation in which an object was isolated from its environment, contained in an apparatus, and physically altered and tested by an experimenter.[5]

In this chapter, I argue that Francis Bacon's concept of experimentation builds on the idea of nature as a creative force, but approaches it as a force that can be captured and put in bonds for the purpose of extracting nature's secrets. As free, Nature is active and creative (*natura naturans*); as placed in bonds (*natura naturata*), Nature is passive and can be studied empirically. Through the "vexations of

84 *Controlling Nature*

art," or the use of technologies to constrain and hold nature, it can be studied, transformed, and improved. In the *Novum Organum* (1620), Bacon wrote, "The secrets of nature reveal themselves more readily under the *vexations of art* than when they go their own way."[6]

During the seventeenth century, a major transformation took place. The origins of both the experimental and mathematical methods that were key to the rise of modern science took shape under natural philosophers, such as Francis Bacon, René Descartes, Thomas Hobbes, Robert Boyle, Baruch Spinoza, Isaac Newton, and Gottfried Wilhelm Leibniz. During the eight decades that intervened between Bacon's publication of *The Advancement of Learning* in 1605 to Newton's *Principia mathematica* in 1687, a new grasp of human power over and prediction of nature arose. Through experimentation, "Nature naturing" would yield "her" secrets; through mathematics, "Nature natured" could be described, predicted, and managed.

Bacon on Nature Naturing and Nature Natured

Francis Bacon rose to power under James I beginning in 1603 when the latter ascended to the throne. He became learned counsel in 1603, attorney general in 1613, privy councilor in 1616, lord keeper in 1617, and lord chancellor and Baron Verulam in 1618. His goals were to extract the secrets of nature for the benefit of humankind. Bacon used the relationship between "nature naturing" and "nature natured" to argue that art (*techne*) could be used to subdue an active "nature naturing" by holding it in constraint and to extract secrets from "nature natured" (see below). Using the new technologies of mining and metallurgy, miners should "search into the bowels of nature" while metallurgists should "shape nature as on an anvil." Bacon's goal was to transform the occult arts of the Renaissance—alchemy, witchcraft, and natural magic—from elite secrets to open knowledge. By the vexations of art, nature would betray her secrets to humanity.[7]

Among other sources available during the Renaissance, Bacon's method may well have been influenced by the Hippocratic treatise *On the Techne* (late fifth century B.C.E.) (see Chapter 1) through the Latin translations of Marcus Fabius Calvus in 1525 and Janus Cornarius in 1538 and 1546. Here the analogy between the human body and nature's body suggests that *techne* (art) could be used to extract the secrets from *physis/natura*. The insights of Heinrich von Staden and Theodor Gomperz (discussed in Chapter 1) on the application of *techne* to nature, causing the body to give up its secrets, apply directly to Bacon's method. The author of *On the Techne* writes:

> When this information is not afforded, and nature herself will yield nothing of her own accord, medicine has found means of compulsion, whereby nature is constrained, without being harmed, to give up her secrets; when these are given up she makes clear to those who know about the art, what course ought to be pursued.

To Gomperz, the language of the Hippocratic treatise *On the Techne* anticipates that of Bacon, in which through art (interrogation via experimentation), nature is forced to release her secrets.[8]

In Book 2, Aphorism 1, *Novum Organum*, (1620) Bacon differentiated between the manipulation of nature to create new entities through the use of power and the study of nature as an active, creating entity to obtain knowledge. The first case (i.e., *natura naturata*) led to power over a passive nature, the second (*natura naturans*) led to knowledge of an active nature. Bacon used the Latin term *natura naturans* (the syntactically correct term being *naturam naturantem*) for the case of knowledge. The goal of knowledge was to discover the underlying "Form" (*Formam*) of nature, by which it created, or "engendered" the things of the natural world.

> On a given body to generate and superinduce a new nature or new natures, is the work and aim of Human Power. [*Super datum corpus novam naturam sive novas naturas generare et superinducere opus et intentio est humanae potentiae.*] Of a given nature to discover the form, or true specific difference, or nature-engendering nature, or source of emanation (for these are the terms which come nearest to a description of the thing), is the work and aim of Human Knowledge. [*Dat autem naturae Formam, sive differentiam veram, sive naturam naturantem, sive fontem emanationis . . . invenire opus et intentio est humanae scientiae.*][9]

In his discussion of Bacon's usage, in his Latin edition of the *Novum Organon* (2nd ed., 1889), Thomas Fowler wrote:

> *Natura naturata* is the actual condition of a given object or quality, or of the aggregate of all objects and qualities, the universe, at any given time; *natura naturans* is the immanent cause of this condition or aggregate of conditions. . . . Hence *natura naturans* is related to *natura naturata* as cause to effect, or again, we may say that *natura naturans* is the active or dynamical, *natura naturata* the passive aspect of nature.[10]

Spedding, Ellis, and Heath, who produced the standard edition of Bacon's *Works* in the 1870s, explained it as follows: "Bacon applies it (*natura naturans*) to the Form, considered as the *causa immanens* of the properties of the Body."[11]

Thus Nature in its active aspect is a free and creative force, *natura naturans*, that can yield knowledge of the natural world—it is the cause of the properties of a particular object. By intervening on the cause (*natura naturans*) Nature can be captured, studied, transformed, and controlled, producing "a new nature or natures" through human power. Bacon's work therefore sets up the possibility for predicting and controlling nature through science (natural knowledge) and technology (art or *techne*) as it would emerge during the so-named Scientific Revolution of the seventeenth century.

Nature, Bacon states in his 1623 Latin revision of *The Advancement of Learning* (*De Dignitate et Augmentis Scientiarum*), exists in three states—at liberty (*libera*), in error (*errores*), and in bonds (*vincula*):

> She is either free and follows her own course of development as in the heavens, in the animal and vegetable creation, and in the general array of the universe;

86   *Controlling Nature*

or she is driven out of her ordinary course by the perverseness (*pravitatibus*), insolence (*insolentiis*), and forwardness of matter (*materiae contumacies*) and violence of impediments (*impedimentorum violentia*), as in the case of monsters (*monstris*); or lastly she is put in constraint (*constringitur*), molded (*fingitur*), and made as it were new by art and the hand of man (*arte et opera humana*), as in things artificial.[12]

While the first and third states of nature may be seen as consonant with *natura naturans* and *natura naturata*, what can be said of the second state, or nature in error? Bacon's second state of nature arises from matter acting perversely and insolently. Here we have an example of the recalcitrance and rebellious audacity of nature described by Neoplatonists, such as Plotinus, but attributed to matter, rather than to the "nous" or "soul" separating joyously from the One to unite with matter. The second state is likewise an example of what Renaissance artists and writers meant by the willful agency of nature, wherein nature acts independently and perversely to create monsters and to render reprisals against improper actions. Bacon's second state of nature is necessary to explain observed discrepancies in the created world that differ from the actions of nature acting in accordance with the laws of nature. This second state of nature would be challenged by Spinoza who argued that if any entities or miracles deviated from the regular and predictable actions of *natura naturans*, it was only because science had not yet discovered the reasons. The laws of nature were totally rational and once understood totally predictable (see Chapter 5).

Bacon and Vexation

Bacon used the term "vexation" (*vexare*) to mean the capture and manipulation of nature for the purpose of obtaining power over the natural world (*natura naturata*). Out of the process of "vexing nature" developed the experimental method of the latter half of the seventeenth century. Vexing nature would lead to power and hence to control. The term, vexation, was present in Bacon's intellectual milieu and, in addition to witchcraft (described earlier), was used in alchemy, the Inquisition, and the *Bible*. Through experimentation (vexation) as an intervention into nature—i.e., the created world of *natura naturata*—nature could be manipulated for the benefit of humankind. The term "vexation" was used in several contexts that were present in Bacon's cultural context that may well have influenced the meaning, or set of meanings, he gave to it. Of prime importance was the process of extracting secrets from nature.

Alchemy

Vexations in alchemy were designed to open the Cabinet of Nature's Secrets. Paracelsus stated that some changes took place by Nature without Art (an idea similar to *natura naturans* and to Bacon's idea of nature at liberty), while others were brought about by Art (*techne*), i.e., by vexation (similar to Bacon's idea that Nature revealed "her" secrets under the "vexations of art"). Although Bacon knew

Vexing Nature 87

and disparaged Paracelsus and alchemy in general, the idea of vexing nature to extract its secrets played a major role in Bacon's philosophy and the development of experimentation in the seventeenth century.

Bacon was intimately familiar with Paracelsus's (1493–1541) *Coelum Philosophorum*, or *Heaven of the Philosophers*, a book that was subtitled, *The Book of Vexations*.[13] Graham Rees has argued that Bacon, while often berating Paracelsus (see the *Temporis Partus Masculus* [ca. 1602]), knew and drew on Paracelsus's cosmology and alchemy in creating a new system of the world—a chemistry of the cosmos—outlined by Bacon in his *Thema Coeli* (1612).[14] Alchemy was considered an Art that would open the Cabinet of Nature's Secrets and alchemical vexations were manipulations of matter in order to transform the base metals into gold and silver. Paracelsus's goal, however, was not the transmutation of metals into gold, but the manipulation of the chemical and spiritual principles, sulphur, mercury, and salt, to create medicines. In his *Coelum Philosophorum*, he wrote, "many secrets are herein contained . . . confirmed by full proof and experimentation." In his *Book Concerning the Tincture of the Philosophers*, he wrote about the "Art" of "transmutations," stating:

> in the fire of Sulphur is a great tincture for gems, which, indeed, exalts them to a loftier degree than Nature by herself could do. But this gradation of metals and gems shall be omitted by me in this place, since I have written sufficiently about it in my Secret of Secrets, in my book on the *Vexations of Alchemists*, and abundantly elsewhere.[15]

In Book I, of *Concerning the Nature of Things*, (1537) Paracelsus wrote: "The generation of all natural things is twofold: one which takes place by Nature without Art, the other which is brought about by Art, that is to say, by Alchemy." Paracelus's ideas concerning Nature and Art bore similarities to Bacon's first and third states of nature as "free and at large" (i.e., *natura naturans*) and as under the "vexations of art" (nature transformed by human power, or *natura naturata*). In the same section Paracelsus also wrote extensively about monsters (Bacon's second state of nature) although Bacon, unlike Paracelsus, attributed monsters to matter acting perversely. Paracelsus's transmutations of the forms of matter likewise bore similarity to Bacon's myth of Proteus as matter changing "herself into divers strange forms and shapes of things." In Book IV, "*Concerning the Transmutations of Natural Objects*," he wrote:

> Transmutation . . . takes place when an object loses its own form and is so changed that it bears no resemblance to its anterior shape, but assumes another guise, another essence, another colour, another virtue, another nature or set of properties.[16]

Although Bacon developed his own theory of the three states of nature and the transformation of matter, he may have drawn some of his inspiration from Paracelsus's treatise, *Concerning the Nature of Things*.

In alchemy, the steps in the transmutation of matter were known as vexations. In transforming natural objects through Art (alchemy), Paracelus wrote, "It is

88   *Controlling Nature*

most necessary to know the steps to transmutation. . . . They are the following: Calcination, Sublimation, Solution, Putrefaction, Distillation, Coagulation, Tincture."[17] In Ben Jonson's play, *The Alchemist* (produced at the court of James I in 1610), the trickster, Subtle, instructs his valet, Face, to "Name the vexations, and the martyrizations / Of metals in the work." Face responds, "Sir, 'putrefaction, solution, ablution, sublimation, cohobation, calcination, ceration, and fixation.'"[18]

Alchemy in its goal of revealing nature's secrets thus provided one context for the idea of vexation as the human manipulation of matter through art. If Bacon's goal was to find a term that represented the transformation of matter under the constraint of technology (art), then alchemical vexation may have been a source for his concept. His use of Proteus, symbol of Matter itself (see his *Wisdom of the Ancients*), was a primary example of vexation.

The *Bible*

Perhaps the most historically significant use of the word *vexation* was in the *Bible*, occurring in the *Old Testament*, Book of "Isaiah." The Catholic version of the *Bible*, the *Vulgate*, in which the Latin appears, states the following verse from Isaiah 28: 19: "*Sola vexatio dabit intellectum auditui*." English translations are: "Only pain shall give understanding." Or, alternatively, "Only tribulation alone will give understanding to the hearing." The King James version, published during the time of Francis Bacon, translates the Latin *vexatio* as vexation: "And it shall be a vexation only to understand the report." Alternatively, "And it shall be a vexation only when he shall make you to understand doctrine." The Latin usage as "pain" and "tribulation" implies that vexation is more than agitation or irritation. The Foxe translation is: "Vexation geueth (giveth) understanding." The phrase, *vexatio dabit intellectum*, was of major significance in the era of the Inquisition. These ideas of vexation were prevalent in Bacon's cultural milieu and in the translation of the *Bible* done for Bacon's patron, King James I of England.

The Inquisition

According to Christine Caldwell Ames in her book, *Righteous Persecution: Inquisition, Dominicans, and Christianity in the Middle Ages*, the concept, *vexatio dabit intellectum*, lay at the root of the Inquisition's rationale for identifying and reforming heretics within the populace. "The prioritization of confession in the inquisitorial method particularly led to inquisitorial torture."[19] If an accused individual refused to confess during an interrogation, pain administered to the body would redirect the understanding. Sometimes imprisonment as the instrument of vexation would result in confession.[20]

Here, similar to Bacon's idea that "the secrets of nature reveal themselves more readily under the vexations of art," constraint and confinement could be used to extract the secrets of the accused and hence the confession. Beyond imprisonment, the causal relationship between vexation and understanding was present in Inquisitional torture. "Torture's purpose was the heretic's acknowledgment, rather than the creation, of truth; torture 'produced' truth in the literal sense of drawing it out ... as here again vexation led to understanding."[21] By analogy, nature under vexation acknowledges or reveals, but does not create the truth, hence vexation leads to the understanding of nature. According to Ames, "inquisitors sought ... an instructive yet conquering interrogation." Inquisitors had the power of "binding and loosing."[22]

Bacon likewise sought to extract the truth through the inquisition of nature under constraint in order to advance human dominion over it. *Inquisitio*, meaning "investigation" or "inquiry," had its roots in Roman law, but in its ecclesiastical usage put an overseeing authority in the position of investigation and judgment. In a somewhat similar vein, Francis Bacon wrote that the scientist, as authority, must not think that "the inquisition of nature is in any way interdicted nor forbidden."[23]

Vexation in Early Modern Science

For Bacon, the idea of vexation, therefore, had several layers of meaning with respect to physical nature (*natura naturata*). First, following the idea of vexation in alchemy, it meant the manipulation of objects by the scientist to produce new combinations of material things not previously existing in nature. Second, following the idea of vexation in witchcraft, it implied the constraint of nature by human hands. Third, following the idea of vexation in the Inquisition, it encompassed the extraction of truth under duress, or nature subject to and as revealed by the human mode of questioning. These layers of meaning separate Baconian science from Aristotelian and medieval science in the creation of artificial objects (objects created through art, or *techne*) to be studied through experimental science.

Mary Tiles, following Ian Hacking, has argued that experimental science since Francis Bacon has emphasized the role of intervention in nature through "vexation." Scientists have not sufficiently appreciated "the respect in which modern science intervenes in nature to further its inquiries—the respect in which it relies on, as well as generates, new forms of human artifice."[24] The new outlook depended on a change in the conceptualization of nature, that is, on where to draw the line between the natural and the artificial. Things artificial differ from the things of nature; artifacts are produced by human actions, whereas natural objects are created by the free actions of nature (i.e., *natura naturans*). "In things artificial," Bacon stated, "nature takes orders from man and works under his authority." Artifacts are not natural objects, but, according to Tiles, objects in which "the effects of human action merge with what it is that the investigator is trying to understand."[25]

Drawing on Bacon's idea of vexation, the modern natural sciences, as Hacking elaborated, are built on things that do not occur naturally, such as steam engines, generators, interferometers, cyclotrons, and the like, but nevertheless operate according to nature's laws. Artifacts, created by human hands intervening in nature, function according to the laws of nature and can be studied through experimental science. "Under Bacon's urgings," Tiles points out, "the student of Nature was to participate actively, to force things into previously non-existent configurations in order to see how they behaved; 'the secrets of nature reveal themselves more readily under the vexations of art than when they go their own way.'" Since Bacon, she observes, "Nature came to be regarded as having hidden depths ... of unrealized

90 *Controlling Nature*

potential which needed the manipulation of phenomena, the 'vexations of art' to bring to the surface of actuality."[26]

Baconian science as intervention therefore entails the constraint and transformation of natural entities through technology to the end of creating new natural and artificial forms not previously existing. Beyond material artifacts, such as the generators, voltmeters, and interferometers of the nineteenth century, are the large hadron colliders, genomes, and nano-particles of the twenty-first century. The creation of such novel natural entities through the constraint, manipulation, and vexation of nature lies at the heart of the Baconian vision.

Bacon and Experimentation

Bacon's third state, i.e., nature in bonds, or nature "in constraint, molded, and made as it were new by art and the hand of man," anticipates the confined, controlled experiment of the late seventeenth century. For Bacon, extracting nature's secrets through interrogation is analogous to a judicial trial, in which the subject on the witness stand is forced to answer questions in order to extract the truth ("the inquisition of truth"). One must not think that "the inquisition of nature is in any part interdicted or forbidden," he wrote. Although Nature *per se* cannot speak, it is privy to the facts and knowledge (secrets) to be extracted. Nature must recognize the words of the questions put by the human examiner as written in "her" own language and must in turn give reliable, repeatable answers in that language. By analogy, the scientist designs an experiment in which nature is "put to the question" in a confined, controlled space in which the correct answers (secrets) can be extracted through inquisition.

The *New Atlantis* (written in 1624 and published posthumously in 1627) epitomized Bacon's anticipation of the contained, controlled experiment that emerged by the end of the seventeenth century. In the visionary experiments conducted in Salomon's House, workers contributed to the study, speeding up, and modification of plants, animals, the metals, and the weather for the benefit of humankind. "Laboratories" included perspective houses, engine houses, furnaces, sound houses, mathematical houses, and parks and enclosures. Regarding uncontrollable events such as the weather, Bacon wrote, "We have also great and spacious houses, where we imitate and demonstrate meteors—as snow, hail, rain, some artificial rains of bodies and not of water, thunders, lightnings." "The end of our foundation," he stated, "is the knowledge of causes, and secret motions of things; and the enlarging of the bounds of human empire, to the effecting of all things possible."[27]

During the seventeenth century, experimentalists, inspired by Francis Bacon, began to reveal the secrets of nature that would lead to knowledge of the natural world (*natura naturata*) and to the possibility of human prediction and control.

Natural Knowledge

The idea of natural knowledge, or "knowing nature" as knowledge of the natural world, developed with the founding of scientific societies. Inspired by Bacon's call for an "inquisition of nature" by penetrating into the secret recesses of things, philosophers founded experimental societies in Italy, England, and France to set up investigations into the nature of things. The Italian Accademia del Cimento (1657), the Royal Society of London (1662), and the Paris Academy of Sciences (1666) all emphasized Bacon's experimental method. In 1663, the members of London's "Royal Society for Improving Natural Knowledge" met "to discourse and consider philosophical enquiries" related to "Physick, Anatomy, Geometry, Astronomy, Navigation, Staticks, Magneticks, Chemicks, Mechanicks, and Natural Experiments." The object was to study nature free from the influence of "theology or state affairs." Many of these experiments illustrated the process of vexing nature to reveal its secrets. Especially relevant are experiments done on living things.[28]

Robert Boyle's *New Experiments Physico-Mechanicall, Touching the Spring of the Air...* (1660) used a sophisticated form of Otto von Guerike's air pump to put living

Figure 4.3 Robert Boyle (1627–1691).
From *Old England's Worthies*, London: C. Cox, 1847, p. 168 (illustration), pp. 178–180 (biography). Courtesy of the Wellcome Library, London.

Figure 4.4 Robert Boyle's Air Pump.
From Robert Boyle, *The Works of the Honorable Robert Boyle in Five Volumes*, ed. Thomas Birch (London: A. Miller, 1744), vol. 1, title page graphic by Hubert-François Gravelot.
Inscription reads: "To Know the Supreme Cause from the Causes of things."

things into a chamber from which air could be evacuated to see if they lived or died (especially experiments 40 and 41). In experiment 40 he put several types of insects in the air chamber and then observed their behavior when the air was evacuated. A fly walking on side of the chamber dropped to the floor of the receiver when the air was evacuated. A bee fell off a flower and a white butterfly fluttered up and down and then fell trembling to the floor of the chamber when the pressure was reduced. In experiment 41, Boyle introduced birds. A lark in good health jumped up and down until the bell jar was evacuated after which it appeared sick and entered into violent convulsions. A sparrow lasted seven minutes until it succumbed, but was revived when air was reintroduced. Upon a second evacuation it fell dead. The same behavior was observed when a mouse was placed in the chamber and the air removed. It appeared to be giddy, staggering about until it finally fell unconscious.[29]

In 1667, Thomas Sprat reported on additional experiments done by Royal Society members on living creatures (such as chickens, snakes, frogs, and fish) in the rarified air of the air pump, on injecting dogs with liquid infusions, and on blood transfusions. Other natural philosophers transfused blood between dogs, dogs and sheep, and even between sheep and humans to see the results. Experimentalism in a confined space exemplified Bacon's idea of vexing nature to induce it to reveal its secrets. In this way *natura naturans* as freely creating nature could be understood, controlled, and managed. But *natura naturans* nevertheless had an unruly side that could not be so confined and controlled. That side was most clearly evident in the problem of the weather, a problem that engaged Boyle's attention, instruments, and observations.[30]

Figure 4.5 An Experiment on a Bird in the Air Pump, 1768.
Painting by Joseph Wright of Derby, 1768, courtesy of The National Gallery, London.
In this painting, a pet cockatoo has been removed from a cage (shown in the upper right corner) and placed in an air pump from which the air is evacuated. The experimenter, whose hand is placed near the stopcock, holds the power to halt the evacuation and return air to the jar to revive the bird. The women are stereotypically emotional, looking in horror at the air pump, hiding their eyes or looking at the men, thereby experiencing the results vicariously. The men control the outcome via the stopcock, stare directly at the experiment with open curiosity, or contemplate the larger philosophical meaning of death.

Unruly Nature

Boyle's experimental work using the barometer and air pump led to the effort to understand the movement of the ambient air and its larger relationship to the weather. Despite the uniformity of the laws of nature, Boyle noted, there existed "some seeming irregularities (such as earthquakes, floods, famines, &c.)." Indeed despite all the successes of seventeenth-century science in celestial and terrestrial mechanics, the weather remained an intractable problem. In 1692, Boyle published his *General History of the Air* in which he advocated that people keep diaries of day-to-day weather observations.[31]

The weather exemplified nature's most unruly aspect. Unanticipated storms, floods, and droughts made the harvesting of food and fuel problematic and the consequences for human health and disease unpredictable. Efforts to understand and determine causal relationships affecting weather patterns were undertaken by a number of the century's foremost natural philosophers. The science of meteorology developed from distinctions among different types of

94 *Controlling Nature*

atmospheric phenomena. Aerial meteors pertained to winds; watery meteors to rain, snow, hail, and dew; luminous meteors to rainbows, halos, and aurora; and fiery meteors to comets and shooting stars. Francis Bacon, in his *New Atlantis* (1627), had implicitly referred to some types of watery meteors when he envisioned a laboratory "where we imitate and demonstrate meteors—as snow, hail, rain, some artificial rains of bodies and not of water, thunders, lightnings." He also included a "History of Winds" as one of the natural histories proposed in his 1620 *Historia naturalis et experimentalis de Ventis* (in *Preparative Toward a Natural and Experimental History*). Here he listed experiments, observations, and possible causes of the power of winds and their often violent and extraordinary manifestations.[32]

Interest in the weather and climate change rose dramatically during the seventeenth century. In his book, *Global Crisis: War, Climate Change, and Catastrophe in the Seventeenth Century*, Geoffrey Parker notes that "in the mid-seventeenth century, the earth experienced some of the coldest weather recorded in over a millennium. Perhaps one-third of the human population died." In 1627 the wettest year on record during the previous five hundred years was followed by temperatures so cold that crops died without even ripening. In 1649 the River Thames prophetically froze over following the execution of Charles I and the rest of Europe experienced a seemingly unending winter lasting six months. Between 1666 and 1679 nine summers were exceptionally cold with a series of bad harvests ensuing. So many disastrous harvests and extreme temperatures occurred during the 1690s that climatologists dubbed the century "the Little Ice Age."[33]

In conjunction with extreme weather conditions and the beginnings of systematic record keeping, meteorology as a new form of natural knowledge began to develop. Numerous individuals sent in a variety of records to academies of science and magazines throughout Europe. A model format for keeping weather records was developed by Robert Hooke in 1663. Edmund Halley in 1686 urged sailors everywhere to record changes in the winds. And John Locke kept observations for the year 1692 with explicit instructions about what should be included in day-to-day records. His own tables included readings from his "thermoscope, baroscope and hygroscope" with notes on wind directions, cloud formations, and weather patterns, along with written accounts of "remarkable meteors," such as blizzards, whirlwinds, and auroras. Weathervanes, thermometers, barometers, and hygrometers lent quantification to the developing science, but construction, liquids (mercury or alcohol), and scales varied widely.[34]

But hope that such records would ultimately reveal the secrets of nature, leading to control over the weather, voyages, harvests, and diseases, proved elusive. The complexities of weather, its often unpredictable chaotic patterns, and the irregularity of events and impacts paled in comparison to the predictability of falling bodies, gravitation, and planetary motions. What made mechanics so successful was its ability to abstract away the irregularities of air resistance and friction, reducing matter to ideal spheres, vacuums, and point sources of force. As Lorraine Daston in her article on "Unruly Weather," concludes, "This path, pursued so triumphantly in seventeenth and eighteenth century mechanics, remained blocked to investigators of the weather and many other natural phenomena of perplexing variability. . . . " Or, as Jan Golinski puts it in *British Weather and the Climate of Enlightenment*, "How are atmospheric phenomena to be explained? Does weather exhibit regular patterns over the long run or will it always remain unpredictable?"[35]

The weather and its predictability and control represented the recalcitrance of autonomous nature to be reduced to rational order. The weather could not be confined in an experimental space as Bacon had hoped. Nor could it be reduced to a set of mathematical equations that could be used for prediction and management. It continued to exemplify the unruliness of nature.

Natural Law and the Laws of Nature

Natural law and the laws of nature both stem from the concept of nature as a rational order. In general, natural laws are prescriptive, while the laws of nature are predictive. Natural law refers to a tradition in which the laws of society, hence the laws of justice, are seen to be grounded in nature and prescribe how humans should act. The laws of nature, on the other hand, are descriptions of mathematical regularities that tell us what to expect in particular situations and therefore allow us to predict what will happen in the future. If nature is a rational order, then the cosmos as a whole is a rationally organized system of laws, beliefs, and behaviors. Both the laws of human society (natural laws) and the laws of nature (scientific laws) are rational systems stemming from the underlying rationality of the cosmos. Both use the term "law," the first as the laws that govern human society, the second as the laws that govern nature.[36]

Although they would later evolve in less deterministic ways, by the end of the seventeenth century a new relationship had emerged that linked the laws of nature in natural philosophy to natural law in society, both based on rationality. The unruliness of nature violated the assumption of the rationality of the laws of nature. The unruliness of society violated the assumption of the rationality of natural law. Disorder in the cosmos—the willful dimension of *natura naturans*, as the rebellious audacity of the World Soul or nature defying the will of God—could be suppressed by the rational understanding of the laws of nature. Disorder in society could be suppressed by the rationality of natural law theory. Prediction and prescription come together in the Scientific Revolution as a system of Enlightenment. The rational understanding of nature as active and creative—Francis Bacon's first state of nature—led to knowledge. The errors of nature—Bacon's second state of nature—that produced monsters, freaks, and curiosities when matter acted perversely and insolently, could likewise be explained by understanding the underlying laws of nature. Such an understanding could *potentially* allow for predicting and controlling the unruly and chaotic side of *natura naturans*. Finally, Bacon's third state of nature—nature in bonds—led to power over nature (*natura*

96 *Controlling Nature*

naturata) gained through experimentation and the application of the laws of nature to the improvement of human life.

During the seventeenth century, both natural law and the laws of nature contributed to a new sense of order in society and the cosmos. Following the wars of religion, the Protestant Reformation, and the English Civil War, new theories of social order were put forward by Thomas Hobbes and John Locke grounded in ideas of natural law as a rational system going back to Cicero and Aquinas. For Cicero, the cosmos was rational and purposeful and natural law imposed a moral obligation to contribute to the good of society. In his *De Legibus*, Cicero wrote:

> Law is the highest reason, implanted in Nature, which commands what ought to be done and forbids the opposite. This reason, when firmly fixed and fully developed in the human mind, is Law. . . . Law is a natural force; it is the mind and reason of the intelligent man, the standard by which Justice and Injustice are measured.

For Aquinas, God's eternal law governs the created world; natural law is discovered by reason and is the means by which humans participate in eternal law. Natural law stemming from God was the standard by which all laws in society should be judged. Through reason and conscience, humans could distinguish good from evil and were morally obligated to pursue the good. It was the obligation of the state, in accordance with natural law, to ensure the greatest happiness of all its citizens.[37]

Thomas Hobbes and John Locke saw civil society as grounded in the rationality of natural law, but emerging out of the "state of nature." For Hobbes (1651), natural law was "a precept, or general rule, found out by reason, by which a man is forbidden to do that which is destructive of his life or takes away the means of preserving the same. . . . " The state of nature was one in which people were in constant fear due to the potential for violence and war. To achieve social order, people gave up the right to kill each other and embraced law in the form of a sovereign. The sovereign maintained peace with a sword in one hand to ward off outside enemies and a scepter in the other to maintain internal order. For Locke (1690), natural law protected people's natural rights to life, liberty, and property.

> The state of nature has a law to govern it which obligates every one; and reason which is that law, teaches all mankind . . . that, being all equal and independent, no one ought to harm another in his life, health, liberty, or possessions. . . .

People lived in the "state of nature" until they rationally and voluntarily formed a community governed by a set of laws. The theories of Hobbes and Locke were based on voluntary contracts made between people and a state governed by natural law grounded in reason. If the state failed to keep its contract, people could overthrow it and establish a new government.[38]

During the same period, owing to the theories of Nicholas Copernicus, Johannes Kepler, Galileo Galilei, and Isaac Newton (among others), the older earth-centered cosmos of Ptolemy and its Aristotelian theories of motion were replaced by a new sun-centered cosmos governed by the laws of mechanics and gravitation. A new view of God as a clockmaker, mathematician, and engineer helped to restore a sense of law and order to nature.[39] A rational, law-governed society in which people voluntarily obeyed human laws and a rational, law-governed cosmos that obeyed the laws of nature worked hand in hand to re-create order and optimism.

Conclusion

By the end of the seventeenth century an empirical science based on the constraint (vexation) of nature had emerged. Nature as creative force (*natura naturans*) underwent examination in early modern scientific societies, leading to the idea of natural knowledge and hence to the possibility of control over active nature. Experimental knowledge, developed by numerous natural philosophers, became one pillar of an emerging mechanistic worldview. Observations and experiments on corporeal bodies (*natura naturata*) from the microscopic to the cosmic set out a new form of natural knowledge. The second pillar was mathematics, based on a revival of Euclidian geometry and the development of analytic geometry and calculus. Together these two approaches would provide both power and knowledge leading to the possibility of prediction and control over a nature that could seemingly be willful, recalcitrant, and unpredictable. Natural law and the laws of nature based on human reason restored law and order to a society disrupted by the wars of religion and the breakdown of the Ptolemaic worldview. In the last quarter of the seventeenth century, Baruch Spinoza's philosophy would incorporate *natura naturans* and *natura naturata* into a system of rational order based on natural law.

Notes

1 John Swan, *A True and Brief Report, of Mary Glovers Vexation, and of Her Deliverance by the Means of Fasting and Prayer* (1603), pp. 1–71 separate pagination, in *Witchcraft and Hysteria in Elizabethan London: Edward Jorden and the Mary Glover Case*, ed. with an introduction by Michael MacDonald (London: Tavistock/Routledge, 1991), introduction, pp. ix–xiii, quotation on p. xi.
2 Swan, *True and Brief Report*, in MacDonald, *Witchcraft and Hysteria*, separate pagination, see pp. 45–47, quotations on pp. 45, 47. Swan continues:

> another preacher began to pray and having a little while continued the same, the maid did fall down suddenly into the chair where she remained without motion, her head hanging downward, somewhat inclining towards the shoulder, her face and color deadly, her mouth and eyes shut, and her body stiff and senseless, so as there were that thought, and I think we all might have said, *behold she is dead*. . . . After she had continued a while in this deadly state: suddenly in a moment, life came into her whole body, her mouth and eyes opened, and then lifting up her hands and stretching them wide asunder as high as she could reach, the first word she uttered was, "he is come, he is come" . . . and then on such as stood on each side of her, "the comforter is come, O Lord thou has delivered me." (pp. 46–47.)

3 Writings on demonic vexation in addition to John Swan's *True and Brief Report of Mary Glover's Vexation* (1603) include John Darrel, *A True Narration of the Strange*

98   *Controlling Nature*

and Grevous Vexation by the Devil, of 7 Persons in Lancashire, and William Somers of Nottingham (England [?]: English Secret Press [?], 1600); Anonymous, *A True and Fearefull Vexation of One Alexander Nyndge: Being Most Horribly Tormented with the Deuill, From the 20. Day of January, to the 23. of Iuly. At Lyering Well in Suffocke: with his prayer afer his deliuerance. Written by his owne brother Edvvard Nyndge Master of Arts, with the names of the witnesses that were at his vexation* (London: W. Stansby[?], 1615).

4   Francis Bacon, *The Advancement of Learning* (1605), in *Works*, ed. James B. Spedding, Robert Leslie Ellis, and Douglas Devon Heath, 14 vols. (London: Longmans Green, 1868–1901), vol. 3, p. 333.

5   Bacon, *Wisdom of the Ancients* (1609), in *Works*, ed. Spedding et al., "Proteus," vol. 6, pp. 725–726. On the meaning of vexation, see Carolyn Merchant, "Francis Bacon and the 'Vexations of Art': Experimentation as Intervention," *British Journal for the History of Science* 46, no 4 (2013): 551–599.

6   Bacon, *Novum Organum* (1620), in *Works*, ed. Spedding et al., vol. 4, Bk. I, Aphorism 98, p. 95, italics added.

7   Bacon, *De Dignitate et Augmentis Scientiarum* (1623), in *Works*, ed. Spedding et al., vol. 4, p. 287, 343.

8   Hippocrates, *The Art*, in *Hippocrates*, ed. and trans. W.H.S. Jones (London: William Heinemann, Loeb Classical Library, 1959), vol. 2, quotation on p. 215; Heinrich von Staden, "*Physis* and *Techne* in Greek Medicine," in *The Artificial and the Natural: An Evolving Polarity*, ed. Bernadette Bensaude-Vincent and William R. Newman (Cambridge, MA: MIT Press, 2007), pp. 21–49, see esp. pp. 26, 30–32; Theodor Gomperz, *Die Apologie der Heilkunst: Hippocratis de Arte* (Leipzig: Veit & Comp. 1910), pp. 140–141. I thank Heinrich von Staden for these references.

9   Bacon, *Novum Organum*, in *Works*, ed. Spedding et al., vol. 1, Bk. II, Aphorism I, p. 227 (Latin): "Super datum corpus novam naturam sive novas naturas generare et superinducere opus et intentio est humanae potentiae. Dat autem naturae Formam, sive differentiam veram, sive naturam naturantem, sive fontem emanationis . . . invenire opus et intentio est humanae scientiae"; vol. 4, Bk. II, Aphorism I, p. 150 (English); Henry A. Lucks, "*Natura Naturans—Natura Naturata*," *The New Scholasticism* 9, no. 1 (1935): 1–24, see pp. 8–9.

10 Bacon, *Novum Organum*, ed. Thomas Fowler, 2nd ed. (Oxford: Clarendon Press, 1889), Bk. II, Aphorism I; Lucks, "*Natura Naturans—Natura Naturata*," p. 9; John McPeek, *Bruno, Spinoza, and the Terms* Natura Naturans *and* Natura Naturata (New Orleans: Tulane University, 1965), p. 42.

11 Bacon, *Novum Organum*, in *Works*, ed. Spedding et al., vol. 1, p. 227, note 1.

12 Bacon, *De Augmentis* (1623), in *Works*, ed. Spedding et al., vol. 4, p. 294.

13 Paracelsus, *The Coelum Philosophorum, or Book of Vexations*, in *The Hermetic and Alchemical Writings of Aureolus Philippus Theophrastus Bombast, of Hohenheim, called Paracelsus the Great . . .*, ed. and trans. Arthur Edward Waite, 2 vols. (Berkeley, CA: Shambala Books, 1976), vol. 1, pp. 5–20. For a commentary of the period, see also, Johann Rudolph Glauber, *Commentary on Paracelsus, Heaven of the Philosophers or Book of Vexation*. Glauber wrote: "And of this I assure myself that in the last times, God will raise up some to whom He will open the Cabinet of Nature's Secrets. . . . " John Dee likewise referred to Paracelsus's *Book of Vexations of Philosophers* (http://www.rexresearch.com/alchemy2/dee.htm):

> The fourth is the manner of making Mineral Amber, of which Paracelsus hath only writ in his *Book of Vexations of Philosophers* and in the last edition of his work in the sixth book of his Archidoxes; but because they cannot be made without the help of the Elixirs, therefore they deserve a place among the Elixirs, where I shall discover the virtue or rather the vice of making Amber;

Also:

> Whence Paracelsus, a worthy Master in Magic, seeing fully the nature and the utility of Alchemy, commanding to make the Elixir thereof, when as its natural body cannot anywhere be had, in his *Book of the Vexations of Philosophers* and the sixth of his Magical Archidoxes, teacheth to compound an Artificial Electrum that the Elixir must be made thereof. . . .

14 Graham Rees, "Francis Bacon's Semi-Paracelsian Cosmology," *Ambix* 12, Part 2 (July 1975): 81–101, see pp. 82, 85. Bacon rejected the Paracelsian principle salt, Bacon, *Thema Coeli* in *works*, ed. spedding eta. vol. 5, pp. 545–559.

15 Paracelsus, *The Book Concerning the Tincture of the Philosophers* . . . Transcribed by Dusan Djordjevic Mileusnic from *Paracelsus his Archidoxis: Comprised in Ten Books, Disclosing the Genuine way of making Quintessences, Arcanums, Magisteries, Elixirs, &c. Together with his Books Of Renovation & Restauration* (London: J. H. Oxon, 1660), Bk. IV, final sentences.

16 Paracelsus, *Hermetic Writings*, op. cit., trans. Waite, vol. 1, quotations on pp. 121, 151.

17 Paracelsus, *Hermetic Writings*, op. cit., trans. Waite, vol. 1, quotation on pp. 151.

18 Ben Jonson, "Mercury Vindicated from the Alchemists at Court," *Works*, 11 vols. (Oxford: Clarendon, 1954–1965), vol. 7, pp. 407–417; Ben Jonson, *The Alchemist*, edited with an introduction and notes by Charles Montgomery Hathaway, Jr., Yale Studies in English, Albert S. Cook, editor (New York: Henry Holt, 1903), see pp. 159–160 and p. 305, note 594: "Paracelsus has a treatise entitled *Coelum Philosophorum or Book of Vexations*, Waite's tr. vol. I, p. 1." See also, P. Ball, "Alchemical Culture and Poetry in Early Modern England," *Interdisciplinary Reviews* 31, no. 1 (2006), p. 12 and Charles John Samuel Thompson, *Alchemy and Alchemists* (Mineola, NY: Dover, 2002 [1932]), p. 199.

19 Christine Caldwell Ames, *Righteous Persecution: Inquisition, Dominicans, and Christianity in the Middle Ages* (Philadelphia: University of Pennsylvania Press, 2009), p. 165.

20 Ames, *Righteous Persecution*, p. 165.

21 Ames, *Righteous Persecution*, p. 167.

22 Ames, *Righteous Persecution*, p. 169.

23 Ames, *Righteous Persecution*, p. 2; Francis Bacon, "Preface" to *The Great Instauration* (1620), in *Works*, vol. 4, p. 20.

24 Mary Tiles, "Experiment as Intervention," *British Journal for the Philosophy of Science* 44 (1993): 463–475, see p. 463; Ian Hacking, *Representing and Intervening* (Cambridge, UK: Cambridge University Press, 1983).

25 Tiles, "Experiment as Intervention," quotation on p. 469; Bacon, *Parasceve* (1620), in *Works*, ed. Spedding et al., vol. 4, p. 253.

26 Tiles, "Experiment as Intervention," quotations on pp. 466, 468.

27 Bacon, *New Atlantis* (1627, written 1624) in *Works*, ed. Spedding et al., vol. 3, 129–166, see p. 156.

28 John Wallis (1616–1703), "A Defence of the Royal Society, and the Philosophical Transactions, particularly those of July 1670 . . . in a Letter to the Right Honourable William Lord Viscount Brouncker, March 6, 1678" (London: Thomas Moore, 1678), p. 7:

> We barred all Discourses of Divinity, of State-Affairs, and of News, (other than what concern'd our business of Philosophy) confining our selves to Philosophical Inquiries, and such as related thereunto; as Physick, Anatomy, Geometry, Astronomy, Navigation, Staticks, Mechanicks, and Natural Experiments.

See also Thomas Henry Huxley, "On the Advisableness of Improving Natural Knowledge," A Lay Sermon delivered in St. Martin's Hall on Sunday, January 7th, 1866, and subsequently published in the "Fortnightly Review," p. 11 (1866), in *Collected Essays*

100 *Controlling Nature*

(New York: D. Appleton, 1896), vol. 1. On the Royal Society and the subsequent history of natural knowledge, see Jan Golinski, *Making Natural Knowledge: Constructivism and the History of Science*, 2nd ed. (Chicago: University of Chicago Press, 2005).

29 John B. West, "Robert Boyle's Landmark Book of 1660 with the First Experiments on Rarified Air," *Journal of Applied Physiology* 98, no. 1 (January 2005): 31–39, see p. 37; Robert Boyle, *New Experiments Physico-Mechanicall, Touching the Spring of the Air, and its effects, made, for the most part, in a new pneumatical engine, etc.* (Oxford: H. Hall, for Tho. Robinson, 1660).

30 Dorothy Stimson, *Scientists and Amateurs: A History of the Royal Society* (New York: Schuman, 1948), pp. 84–86; Steven Shapin and Simon Schaffer, *Leviathan and the Air Pump: Hobbes, Boyle, and the Experimental Life* (Princeton, NJ: Princeton University Press, 1985), pp. 32, 121–122, 346–347; Thomas Sprat, *History of the Royal Society*, ed. Jackson I. Cope and Whitemore Jones (St. Louis: Washington University Press, 1958), pp. 218–219.

31 Lorraine Daston, "Unruly Weather: Natural Law Confronts Natural Variability," in *Natural Law and Laws of Nature in Early Modern Europe*, ed. Lorraine Daston and Michael Stolleis (Burlington, VT: Ashgate, 2008), Ch. 14, pp. 233–248; see p. 237, quotation from Boyle on p. 233; Robert Boyle, *General History of the Air* (London: Awnsham and John Churchill, 1692).

32 Daston, "Unruly Weather," p. 236; *Oxford English Dictionary*, cf. "meteor"; Bacon, *New Atlantis*, op cit.

33 Geoffrey Parker, *Global Crisis: War, Climate Change, and Catastrophe in the Seventeenth Century* (New Haven, CT: Yale University Press, 2013), pp. xv–xxv, 3, quotation on p. xv.

34 Daston, "Unruly Weather," pp. 237–241.

35 Daston, "Unruly Weather," pp. 241–248, quotation on p. 248; Jan Golinski, *British Weather and the Climate of Enlightenment* (Chicago: University of Chicago Press, 2007), p. 2.

36 Daryn Lehoux, "Laws of Nature and Natural Laws," *Studies in History and Philosophy of Science* 37 (2006): 527–549, esp. pp. 527–531.

37 Cicero, *Laws [De Legibus]*, trans. C. W. Keyes, Loeb Classical Library (Cambridge, MA: Harvard University Press, 1970), vol. 16, pp. 297–369, quotation from Bk. I, Ch. V, sec. 18–19, pp. 317, 319; Aquinas, *Summa Theologica*, in *Selected Writings*, ed. and trans. Ralph McInerny (New York: Penguin, 1998), "On Law and Natural Law," IaIIae 90–94, pp. 611–652.

38 Thomas Hobbes, *Leviathan* (Aalen, Germany: Scientia, 1966 [1651]), Part 1, Ch. 14 (quotation); John Locke, *Two Treatises of Government*, ed. Thomas I. Cook (New York: Hafner Press, 1973 [1690]), *Second Treatise*, Ch. 2, "Of the State of Nature," sec. 6 (quotation), Ch. 13, sec. 149.

39 Carolyn Merchant, *The Death of Nature: Women, Ecology, and the Scientific Revolution* (San Francisco: HarperCollins, 1980), pp. 225–226, 288, 290; David Kubrin, "How Sir Isaac Newton Helped Restore Law 'n' Order to the West," *Liberation* 16, no. 10 (Mar. 1972): 32–41.

5 Natural Law

Spinoza on *Natura naturans* and *Natura naturata*

In 1656, a twenty-four-year-old Baruch Spinoza (1632–1677) sat before a panel of judges in a wooden warehouse that served as the Amsterdam synagogue of Jewish descendants expelled from Portugal by the Inquisition. Witnesses testified about his numerous heresies. They alleged that he believed that the books of Moses were a human creation, the soul is mortal, and the corporeal world is part of God.[1]

On July 27, the verdict against him was read:

> The lords of the Mahamad . . . having long known of the evil opinions and deeds of Baruch de Espinoza, have endeavored by various ways and promises to turn him from his evil ways. But . . . daily receiving more information about the abominable heresies which he practiced and taught and about the monstrous deeds he did, . . . have decided . . . that the said Espinoza should be excommunicated and expelled from the people of Israel.[2]

Offered the opportunity to recant, Spinoza declined. On receiving the verdict, he remained serene. Faced with excommunication for his family if they so much as spoke, ate, or even lived with him, he left the Jewish community in Amsterdam. He took up a life of philosophy and freedom in the wider, tolerant Dutch society. He lived outside Amsterdam for several years where he became an expert at the craft of lens-making, before moving to The Hague in 1670. Ultimately, the glass dust from his lens-making profession caused the lung damage that led to his early death in 1677 at the age of forty-five.[3]

In this chapter, I argue that Spinoza developed a rational system of philosophy that militated against chaotic, unruly nature and for the idea of a rational society based on natural law and a rational cosmos described by the laws of nature. I distinguish between natural law as applied to society and the laws of nature as applied to the universe as a whole. For Spinoza, natural laws govern political and moral life; they are prescriptive and normative. The laws of nature, on the other hand, are mathematical and logical; descriptive and deterministic. Through rationality any apparent disorder in nature can be explained away and disorder in society can be held at bay. Both types of law stem from belief that the universe could be understood as a totally rational entity.

Figure 5.1 Baruch Spinoza (1632–1677).
Anonymous, ca. 1665.
Courtesy of *Herzog August Bibliothek*, Germany.

In developing his system, Spinoza rejected earlier beliefs (e.g., Plotinus, Lucretius, Renaissance writers—see Chapter 3) in the chaotic dimension of nature naturing (*natura naturans*), while embracing its rational, creative, active aspect. In so doing, he identified nature (*natura naturans*) with the created world (*natura naturata*) that could be described by logic and mathematics. God and Nature were one (a position later identified as pantheism, see below). Only one Substance, God or Nature (*Deus sive Natura*), existed. That Substance (Being, or reality), manifested itself simultaneously in many Attributes, two of which were available to humans—thought and extension. God was not corporeal (as his accusers had alleged), but there was a corporeal, i.e., "extended," aspect or Attribute of God. God/Nature was active, changing, and in continual process. Through understanding nature and describing it mathematically, its actions could be understood, predicted, and potentially controlled. Heavily influenced by and indebted to Descartes, Spinoza saw his philosophical system as responding to the problems posed by Cartesian dualism.

In contrast to many who preceded him, Spinoza asserted that Nature was not capricious in any way. *Natura naturans* had no rebellious identity or personality, no soul or "will of its own" that caused it to act in defiance of God. Any apparent willfulness within nature was incorrectly projected onto it by humans. Earthquakes, droughts, and famines were not a consequence of nature's reprisals against humans; rather humans irrationally insinuated such ideas onto the will of God/Nature. The laws of nature in the physical world and natural law in the human world converged to create a rational order, with a goal of establishing harmony, order, certainty, and coherence in the cosmos as a whole.

Spinoza's system was a major influence on Gottfried Wilhelm Leibniz (1646–1716) who argued that while God had indeed created a rational world, he remained a personal God (see Chapter 6). Spinoza also influenced a number of philosophers and writers who considered themselves pantheists, along with a number of scientists, including Albert Einstein (1879–1955).

Spinoza's Evolution

Born in 1632, Spinoza was the son of Portuguese Jews who sought freedom from persecution. His father, Michael, was a prosperous trader living in a well-to-do Jewish neighborhood in Amsterdam, where he and his wife Hanna bore five children before Michael died of lung disease when Spinoza was twenty-one. Spinoza studied Hebrew at a Jewish school with a demanding educational program at which he excelled. Amsterdam was a center of world trade, scientific advancement, art, and most important for Spinoza, a place of intellectual and religious freedom. Among its well-known scientists were Christiaan Huygens (1629–1695), inventor of the pendulum clock that would become the symbol of the mechanical philosophy, and Anton von Leeuwenhoek (1632–1723), renowned microscopist who discovered and described numerous micro-organisms. And it was here that Descartes spent much of the last two decades of his life (before dying in Sweden in 1660). A publishing mecca, the city had four hundred bookstores and numerous publishing houses for manuscripts sent from around the world.[4]

It was at one such bookstore that Spinoza met radical free thinker, bookseller, and teacher Frans van den Enden in 1653 with whom, at the age of twenty, he began to study Latin, Greek, and Cartesian philosophy. Under van den Enden's tutelage,

104 *Controlling Nature*

Spinoza's ideas began evolving in more radical directions—toward notions that God/Nature was a Substance, the soul dies at death, and the book of Moses was the word of a mortal man and not of God. During this period he wrote his "Short Treatise on God, Man, and His Well-Being" (discovered only much later in 1810).[5]

In 1663, Spinoza published his first work of philosophy, the *Principia Philosophiae Cartesianae*, an exposition of *Descartes's Principles of Philosophy* (which had appeared in 1644). Two years later he began work on his *Tractatus Theologico-Politicus*, the *TTP* (in English, *The Theological-Political Treatise*), which was published anonymously in 1670, but known to be the work of Spinoza. After his death in 1677, his *Ethica Ordine Geometrico Demonstrata* (the *Ethic*) appeared.[6]

One of those who admired Spinoza's emerging rationalism was a young mathematician named Walther Ehrenfried von Tschirnhaus, who read the *Tractatus* and opened a correspondence with Spinoza on its more subtle points. In 1674 Tschirnhaus visited The Hague where Spinoza was then living and, after conversing with him, gained sufficient respect that Spinoza entrusted him with parts of his manuscript in progress, the "Ethic," along with a promise not to share it with anyone without his express permission. From there, early in 1675, Tschirnhaus traveled to London to visit Henry Oldenburg, Secretary of England's Royal Society, to whom he extolled Spinoza's ideas. Upon leaving, he ultimately reached Paris, carrying with him Spinoza's manuscripts along with a letter of introduction from Oldenburg to Gottfried Wilhelm Leibniz. In Paris, numerous conversations between Tschirnhaus and Leibniz took place over several months concerning mathematics, philosophy, and ultimately, the ideas of Spinoza on the relationship between God and Nature. These conversations illuminated the extent to which Spinoza had already developed his view that God/Nature was a single Substance and that he had rejected the idea of chaos and disorder in nature.[7]

Spinoza on Society

Spinoza wrote during the 1660s and 1670s following a period of social disruption in Europe that included the wars of religion and the English Civil War. His philosophy stresses order in both nature and society founded on rationality and logic. Not only is order a necessary component of nature, it is essential to civil society. Social order results from a society governed by reason, while disorder and anarchy stem from irrationality.

In the *Tractatus Theologico-Politicus* (*TTP*), Spinoza (like Thomas Hobbes in his 1651 *Leviathan*) argued that people in the "state of nature" are prone to violent, disorderly conduct. "The state of nature . . . must be conceived as without either religion or law. . . . " All beings, by nature, are endowed with a striving (*conatus*) to preserve their own lives. "[I]t is the sovereign law and right of nature that each individual should endeavor to preserve itself as it is." In humans, self-preservation entails a self-interest through which people surrender to their "fleshy instincts and emotions" resulting in anger, hatred, and violence. But reason can overcome these instincts. Without reason, people will live in misery succumbing to avarice, ambition, envy, and hatred. True self-preservation requires civil society rather

Natural Law   105

than anarchy. "[N]o society can exist without government, and force and laws to restrain and repress men's desires and immoderate impulses." A democracy is the most natural form of government, Spinoza argued, and the most consonant with the retention of individual liberties. In forming a democracy, people hand over their individual powers to the body politic, the laws of which everyone is bound to obey, but no one hands over so much right that he/she has no voice. The majority of the people thus agree to a rational form of government "under the control of reason, so that they may live in peace and harmony."[8]

Spinoza further elaborated on order in a rational, free, and just civil society in Part IV of his *Ethic*. People make a rational choice to live in a free society in accordance with the "common laws of the state." Rather than living in fear of their fellow men or in fear of punishment and suffering by a larger power, they are guided toward self-preservation through reason. Love overcomes hatred; good surpasses evil; justice prevails over injustice. In a just society, people endeavor to remove hindrances to achieving that society, such as "hatred, anger, envy, derision, [and] pride." Because "all things follow from the necessity of the divine nature," it is only from thinking in a "disturbed, mutilated and confused fashion" that evil and injustice result. The rationality of the laws of nature form a model for rationality in the laws of society. In this way, social disorder can be overcome by order.[9]

Certainty

Of primary importance for seventeenth-century philosophers in resolving the issue of disorder in both nature and society was the problem of certainty. There could be no place for spontaneity in nature or vital forces within matter. Every effect must have a cause. Both Pierre Gassendi (1592–1655) and René Descartes (1596–1660), who preceded Spinoza, were significant figures in dealing with the issue of certainty, although they resolved it in different ways. Gassendi argued that God was not the animate World Soul, but the governor of the world. In the section on Physics in his *Syntagma philosophicum* (published 1658), he argued that the term "nature" was used in two different ways—ways that were similar to *natura naturans and natura naturata*. In the active sense, nature was called *vis ageni* or *natura divina*, the nature that sustains and supports everything. In the passive sense, it was *universitas rerum* or the totality of existing things.[10]

Likewise, Descartes did not use the terms *natura naturans and natura naturata*, but his *Principles of Philosophy* played a major role in the way Spinoza resolved the problems of order and chaos and Descartes's mind-body dualism. In fact, the only work published in Spinoza's own name was titled, *The Principles of Descartes's Philosophy*.[11]

Several problems raised by Descartes influenced Spinoza's philosophy. First, Descartes differentiated between body and mind and the extended and unextended as two substances. Physical beings are extended and unthinking, while psychic beings are thinking and unextended. But how does an event in one produce an event in the other? Descartes's dualism posed a critical problem for Spinoza (and subsequent philosophers), one that Spinoza ultimately resolved by positing

106 *Controlling Nature*

the existence of one Substance with many Attributes, only two of which were available to humans—thought and extension.

Second, Descartes emphasized the role of mathematics (Euclidean and Analytic Geometry) in leading to certainty (hence predictability). In his *Discourse on Method* (1636), Descartes wrote, "Most of all I was delighted with mathematics, because of the certainty of its demonstrations and the evidence of its reasoning. . . . " The idea of the certainty of mathematics (and uncertainty of the senses) influenced Spinoza's rationalism.[12]

Third, Descartes's view that the universe *could* have started from chaos was fundamental to his own ideas on the vortex theory of matter, the plenum, and the relationships among the corpuscles of the material world. The problem of the possibility of an original chaos (which Spinoza rejected) raised the question of the cosmic creation as an eternal order. In *Le Monde* (*The World*, written 1629–1633, published 1644 and 1677), Descartes argued that the world could have originated from chaos as a plenum of corpuscles in vortex motions that were rearranged in accordance with the laws of nature to become the world we know today. He wrote:

> For God has so wondrously established these laws that, even if we suppose that He creates nothing more than what I have said, and even if He does not impose any order or proportion on it but makes of it the most confused and most disordered chaos that the poets could describe, the laws are sufficient to make the parts of that chaos untangle themselves and arrange themselves in such right order that they will have the form of a most perfect world, in which one will be able to see not only light, but also all the other things, both general and particular, that appear in this true world.[13]

Through vortex motion, the planets could have been created to orbit the sun—in accordance with Copernicus's heliocentric hypothesis. Descartes also described how the universe could have arisen from chaos by the laws of nature imposed on female matter:

> But I do not want to delay any longer telling you by what means nature alone could untangle the confusion of the chaos of which I have been speaking, and what the laws of nature are that God has imposed on her. Know, then, first that by "nature" I do not here mean some deity or other sort of imaginary power. Rather, I use that word to signify matter itself, insofar as I consider it taken together with all the qualities that I have attributed to it, and under the condition that God continues to preserve it in the same way that He created it.[14]

Influenced by Descartes's ideas, Spinoza published his own work, *The Principles of Descartes's Philosophy*, in 1663 as a way of explaining Descartes's system to a student he was tutoring. In it, he set out Descartes's dualistic philosophy in the form of definitions, axioms, propositions, demonstrations, scholia, lemmas, and corollaries—a format he would later use in his *Tractatus* and *Ethica*. The book dealt with Descartes's ideas about the mind as thinking substance and body as extended substance; God as the omniscient, incorporeal creator and conserver of all things; and the motion of bodies.[15]

In an Appendix to *The Principles*, Spinoza added his *Cogitata Metaphysica* in which he made his own points about the issues Descartes had discussed. Here he raised the issues of God's eternity, what is created and uncreated, and the problem of chaos. He moved toward his monistic system in which God is a creating substance and one with the created world. In Chapter IX on "The Power of God," he stated that God had existed from eternity and is pure activity. Here he used the term *natura naturata* to refer to the created world. "The *natura naturata* must be considered unitary. Whence it follows that man is a part of Nature." Moreover, there is no chaos. "We will not pause to refute the opinion of those who think that the world as chaos, or a matter devoid of form, is co-eternal with God, and so far independent of Him." The world as a totally rational system in which no miracles or chaotic events existed was further developed in his next work, the *Tractatus Theologico-Politicus*, published anonymously in 1670.[16]

Theology and Nature

Spinoza's *Tractatus Theologico-Politicus* (*TTP*) moved him down a pathway toward ever-increasing rationality, removing the *Bible* from the realm of the mysticism of God's word to a work authored and created by human beings. In the *TTP*, he argued that the *Pentateuch* (or first five books of the *Bible*) was the work of several human authors over many years. Miracles were fantasies and products of wishful thinking. They were impossible for logical and rational reasons. Moreover, prophets had no power to predict the future and the Jews were neither special nor chosen. They were historical peoples living in historical times who encountered and survived extraordinary events. Established religion, Spinoza held, exploited its believers to its own ends, while entrenching hierarchies of clergy. The *TTP* is significant, however, not only for its theology but for its interpretation of nature as a rational system and for its rejection of miracles.[17]

In the *TTP*, in Ch. 6 "On Miracles," Spinoza confronted the issue of apparent miracles. He held that ordinary people viewed any natural phenomenon they did not understand as the work of God. Thus whenever nature failed to follow its ordinary course (as in sudden, violent natural events), it became an argument for God's willful intervention into the natural world. They imagine God to be all powerful, he asserted. "[T]hey style unusual phenomena 'miracles' and prefer to remain in ignorance of natural causes. . . . " They "only admire the power of God when the power of nature is conceived . . . as in subjection to it."[18]

Spinoza's answer to apparent miracles was that God/Nature is a totally rational system. Anything that is apparently done by God's will is actually done according to reason and hence the laws of nature. Thus anything that happens, happens naturally. If something, such as a miracle, could occur that is contrary to the laws of nature, then God would be acting contrary to his own nature, a conclusion that is clearly absurd. "For otherwise what we are saying is that God has created a

108 *Controlling Nature*

nature so impotent and with laws and rules so feeble that He must continually give it a helping hand." If humans cannot explain an event, he went on, then they simply have not yet discovered the law of nature that explains it. As humanity gains greater understanding of the behavior of the natural world, it will gain greater understanding of God.[19]

Spinoza gave a number of examples of miracles from the *Bible* and in each case rejected the biblical explanation and offered a natural cause. If locusts reached the land of Egypt by the apparent command of God, it simply meant that they came on a wind that blew for an entire day and night. If the earth was said to be sterile due to man's sins, then it was clear that a long drought had occurred. If prayers apparently caused rain, the resulting fertility was due to nature, not to God's favorable response. In the *Bible*, explanations were given in poetic terms to appeal to ordinary people. But such irrational explanations have been written into the *Bible* by irresponsible and blasphemous people. Spinoza concluded, "We may, then, be absolutely certain that every event which is truly described in Scripture necessarily happened, like everything else, according to natural laws; and if anything is there set down which can be proved in set terms to contravene the order of nature . . . we must believe it to have been foisted into the sacred writings by irreligious hands; for whatsoever is contrary to nature is also contrary to reason, and whatsoever is contrary to reason is absurd, and, *ipso facto*, to be rejected."[20]

Historian of philosophy Richard Popkin argues that for Spinoza ordinary people are far more affected by manifestations of disorder than by order. "[T]he adherents of popular religion are more impressed by the irregularities in the natural world than by the regularities. Earthquakes, diseases, etc. became the evidence that the gods were angry about some human behavior and had to be placated." For Spinoza, if mathematics had not provided another standard of truth that eliminated final causes and teleological explanations, then God's rationality would have been forever hidden. Most people embrace religion because they ascribe natural disasters, catastrophes, prodigies, and miracles to God. When they discover that "the course of nature is regular and uniform, their whole faith totters and falls to ruins."[21]

Upon publication, Spinoza's *Tractatus Theologico-Politicus* was circulated widely and its anonymous author, whose name was soon known, was branded as a heretic and an atheist. Denounced as vile, impious, and corrosive by the clergy, the underground book found its way into the libraries of the *intelligentsia* across Europe. Some expressed grudging admiration for the attention it garnered; others were captivated by the freedom of thought it boldly asserted.[22]

Nature

From the time of his earliest writings, Spinoza began developing his concept of "God or Nature" (*Deus sive Natura*) as a rational, logical entity that acted from within, as opposed to a voluntarist God who imposed his will on the world from without. In writing about nature as a single, eternal, self-causing substance, *natura naturans*, that was one with God, he drew on historical sources, transforming them to create his own monistic theory. It is not clear where he obtained the terms *natura naturans* and *natura naturata*. In his "Short Treatise" (ca. 1660) he referred to the Thomists as having used the term *natura naturans* to mean God as a "being beyond all substances," which he noted was contrary to his own meaning of a God that was immanent within substance. But while Aquinas had indeed used the concept of *natura naturans* in his *Summa Theologica* (see Chapter 2), he did not use *natura naturata*.[23]

What then were Spinoza's sources? Although he did not refer to his historical antecedents, he apparently knew the ideas of Aristotle, Augustine, and Aquinas and possibly those of Nicholas of Cusa and Giordano Bruno. He may or may not have known of the *natura creans* and *natura creata* of Johannes Scotus Erigena directly, but perhaps absorbed them indirectly as reflected in the work of Aquinas, those whom he called "the Thomists," and/or the scholastics who translated Averroes's commentaries on Aristotle (see Chapter 2).[24] Spinoza's tutor Frans van den Enden, a Catholic freethinking intellectual, knew of and secretly admired Bruno's work, and it was in his company that Spinoza presumably studied the scholastics and read the philosophies of Bruno and Descartes.[25]

It is in *Cause, Principle, and Unity* (1584) that Bruno used the term *natura naturans* associating it with Aristotle's concepts of male form and female matter in conjugal unity as the immanent soul of the world. Benjamin Fuller argues that " . . . Bruno uses the terms *natura naturans* or creative nature and *natura naturata* or created nature. These terms were later adopted by Spinoza to denote essentially the same distinction." John McPeek points out that both Bruno and Spinoza identified God with Nature, both located perfection in the intellectual love of God, and for both men, all of nature was animate. Spinoza seems, at least, to have been influenced by Bruno's pantheistic ideas early in his thought, while later incorporating his differences with Descartes. Bruno's pantheistic monism, which contrasted sharply with Descartes's dualism, presented him with ontological choices as he developed his ideas on the nature of God, substance, infinity, and eternity.[26]

Natura Naturans and *Natura Naturata*

It was in the context of his emerging pantheistic monism that Spinoza attempted to resolve some of the deepest tensions surrounding the problem of "creating" and "created" nature. Rather than seeing the active, creating force (*natura naturans*) as located in a transcendent God, or a Neoplatonic *Nous*, above and separate from Nature, he argued that God was one with Nature. Rather than seeing God as creating the universe at some definitive point in time, *ex nihilo*, God/Nature was an eternal, self-causing, on-going single substance. Nature (*natura naturans*) was thus self-acting, uncreated, and identical with God.

Similarly, rather than seeing a separate, created world that resulted from the activity of God/Nature, Spinoza considered *natura naturata* to be the totality of the beings produced by and resulting from *natura naturans*. Thus the phenomenal or created world was the product of the active, creating world. This phenomenal world was the world of creatures dependent on God and created by Him. Instead

110   *Controlling Nature*

of endowing God or Nature with a will, Spinoza argued that Nature developed by and from its own internal logic. That logic existed within Nature itself as law and was manifested as the Laws of Nature.

Spinoza discussed the relationships between the creating and created worlds (*natura naturans and natura naturata*) and those between natural law and the laws of nature in two main places. In his first philosophical work, the *Short Treatise on God, Man, and His Well-Being* (written ca. 1660 and circulated among friends, but not published until the mid-1800s), he wrote, in Chapter 8:

> [W]e shall briefly divide the whole of Nature—namely, into *Natura naturans* and *Natura naturata*. By *Natura naturans* we understand a being that we conceive clearly and distinctly through itself, and without needing anything beside itself (like all the attributes which we have so far described), that is, God. The Thomists likewise understand God by it, but their *Natura naturans* was a being (so they called it) beyond all substances.
>
> The *Natura naturata* we shall divide into two, a general and a particular. The *general* consists of all the modes which depend immediately on God, of which we shall treat in the following chapter; the *particular* consists of all the particular things which are produced by the general mode. So that the *Natura naturata* requires some substance in order to be well understood.[27]

The things, or particulars, that are in God, or in God's attributes (as Spinoza expressed it in the *Ethic*, see below) are called modes. In the "following chapter" (i.e., Ch. 9), "On *Natura Naturata*," he elaborated:

> Now, as regards the *general Natura naturata*, or the modes, or creations which depend on, or have been created by, God immediately, of these we know no more than two, namely, *motion* in matter, and the *understanding* in the thinking thing. These, then, we say, have been from all eternity, and to all eternity will remain immutable. A work truly as great as becomes the greatness of the work-master.[28]

In 1663, in the Appendix to his *Principles of Descartes's Philosophy*, Spinoza also used the term *natura naturata* in a limited way (as discussed above): "The *natura naturata* must be considered unitary. Whence it follows that man is a part of Nature."[29]

Then, in his *Ethic* (1677), he further differentiated between *natura naturans* (nature naturing, or nature in the active sense) and *natura naturata* (nature natured, or nature already created). Nature was synonymous with God. *Natura naturans* was God as a free cause, an eternal, infinite essence. *Natura naturata* (the natural world) was that which followed from God's own nature by necessity.[30]

In the Scholium to Part I, Proposition 29, of the *Ethic*, he wrote:

> Before I go any farther, I wish here to explain, or rather to recall to recollection, what we mean by *natura naturans* and what by *natura naturata*. For,

from what has gone before, I think it is plain that by *natura naturans* we are to understand that which is in itself and is conceived through itself, or those attributes of substance which express eternal and infinite essence, that is to say, God in so far as He is considered as a free cause. But by *natura naturata* I understand everything which follows from the necessity of the nature of God, or of any one of God's attributes, that is to say, all the modes of God's attributes in so far as they are considered as things which are in God, and which without God can neither be nor can be conceived.[31]

The universe itself emerged by necessity from the laws of logic. The interrelationship between *natura naturans* and *natura naturata* and hence the natural world was thus an expression of an inner logic.

Natural events, therefore, happened according to the eternal order. They expressed regularities in nature and defined and limited what was possible. In this respect, natural law and the laws of nature were synonymous. Alfred Weber writes: "The words God and universe designate one and the same thing: Nature which is both the source of all beings (*natura naturans sive Deus*) and the totality of these beings considered as its effects (*natura naturata*)." God/Nature as substance has "neither intellect nor will: for both presuppose personality." God is not the external "'author of nature', but nature itself." God as substance is absolutely free. "Absolute freedom" states Weber, "excludes both constraint and caprice." Thus, as a consequence of the very meaning of *natura naturans* within Spinoza's system, chaos and disorder are excluded.[32]

How do the concepts of *natura naturans* and *natura naturata* relate to Substance and its modes, i.e., the things of the perceived world? Spinoza distinguished between Substance and its modes in Letter 12 written to Ludovicus Meyer on April, 20, 1663 (without using the terms *natura naturans and naturata*). Substance is the "whole order of nature." Substance (Being) is one, indivisible, infinite, and eternal (i.e., *natura naturans*). It can be understood only by reason. Its very existence follows from its essence. The perceived world (i.e., *natura naturata*) comprises modes or states of Substance. Trees, humans, and chairs are modes of the Attribute extension; thoughts, passions, and sensations are modes of the Attribute thought. While these exist and have essences, they do not have to exist. Their existence does not follow from their essence and we cannot predict at what point they will still exist in the future. Although some modes are infinite and eternal (through their attribute), in general, modes exist in time. Thus trees and people are born and die; chairs can be crafted and destroyed.[33]

These individual things (modes or states of Substance) are not Substance itself. They are individual parts, but those parts taken together do not compose the "whole of nature." The parts (modes) are extended—they take up space—but in and of themselves they do not constitute Substance. Philosophers who argue that Substance is made of corpuscles or atoms that are extended are simply wrong, because Substance is indivisible, infinite, and is not composed of extended parts. "Those who think that Extended Substance consists of parts, or bodies really distinct from one another, are talking foolishly, not to say madly. . . . [T]he whole

112 *Controlling Nature*

medley of arguments by which Philosophers generally try to show that Extended Substance is finite, collapses of it own accord: for they all suppose that corporeal substance is composed of parts." Spinoza, therewith, categorically rejected the arguments of seventeenth-century natural philosophers who had tried to revive the ideas of atomists such as Democritus, Epicurus, and Lucretius.[34]

Natural Law

In his *Tractatus Theologico-Politicus* (*Theological-Political Treatise*), Spinoza moved from a discussion of nature itself to "natural law" as applied to society. Here he distinguished "natural law" from "the laws of nature." There are two types of laws, deriving either from "natural necessity or human decision." The first type of law, Divine law, stems from "the natural light of reason" and arises by "natural necessity" out of God's logic. These "laws of nature" govern the created world (*natura naturata*) and follow from the "very nature or definition of a thing." They are descriptive rather than prescriptive; deterministic, rather than normative. One example of the "laws of nature," for Spinoza, is Descartes's law of motion (quantity of motion, $m|v|$), or what later, as mv, came to be known as momentum: "When one body strikes a smaller body, it only loses as much of its own motion as it communicates to the other." Another example is, "when a man recalls one thing he immediately remembers another which is similar. . . . "[35]

The second type of law applies to society. Natural laws pertain to political and moral affairs. They help to guide human actions. They are normative and prescriptive, rather than deterministic and descriptive. These are the laws that "men prescribe to themselves and to others in order to achieve a better and safer life, or for other reasons." Similar to the argument in Hobbes's 1651 *Leviathan*, Spinoza wrote that "men give up their right which they receive from nature . . . and commit themselves to a particular rule of life." This second type of law depends on human decision making. Humans are not cooperative by nature, but by adhering to normative laws their decisions and actions can be guided toward a rule-governed, social order. Humans acting in accordance with natural law can achieve their greatest good.[36]

Presciently, for the future of democratic societies, Spinoza argued that the state does not exist to serve God, but instead to serve the people. It does not reflect the divine order, but is and should be a government of the people. Its purpose is freedom. If it does not fulfill its purpose, it can be changed by the collective will.[37]

Mathematics

Central to the emergence of rationalism in nature and society, and ultimately to the mechanistic view of nature, was the role played by mathematics in the quantification of nature. Spinoza's method arises from Euclidian geometry, revived in the seventeenth century and used by Galileo, Descartes, and Newton, among others, as the method of proof. Propositions, axioms, corollaries, and geometric diagrams constitute the means of demonstrating the laws of nature. Spinoza, however, uses geometry as a metaphysical method for presenting his system of pantheistic monism. While geometry is his method of proof in demonstrating the existence of Substance (*natura naturans*), mathematics as quantification is useful in understanding the created world of phenomena (*natura naturata*).

For Spinoza, it is only through reason that Substance (i.e., *natura naturans*, creating nature) can be apprehended. The created world (i.e., *natura naturata*), the world of everyday phenomena as perceived by the senses, on the other hand, must be grasped through the imagination. Mathematics is an aid and a corrective to the imagination in understanding the motions and behaviors of individual beings (modes, such as trees, people, chairs, etc.). Quantity can be considered both through reason and through the imagination (with the help of the senses). Considered through reason, quantity is indivisible, infinite, and one. Considered through the imagination, it is divisible, finite, and composed of parts. Mathematics therefore is useful for understanding the modes (parts) of the everyday world.[38]

In the everyday world of trees, people, and chairs (modes of Substance), duration is characterized by time (minutes, days, years) and quantity by number (inches, feet, miles). Time and number help the imagination to understand the everyday world. But they do not help us to understand Substance (i.e., "the course of nature"); that can only be grasped by the intellect through reason. Mathematics aids the imagination in understanding the everyday world. Time and number can be added, subtracted, divided, and multiplied. Time and number can thus be used to characterize the extended world of everyday life. Mathematics is the key to understanding the world of phenomena (i.e., *natura naturata*). Reason is the key to understanding the world of Substance (i.e., *natura naturans*).[39]

The highest form of love for Spinoza is the intellectual love of God. Because God is consonant with nature (*Deus sive Natura*), this means the intellectual love of nature itself as *natura naturans*. But Spinoza's pantheistic monism and his rejection of atomistic parts put him at odds with much of mainstream mechanistic theory. By the late seventeenth century, drawing on ideas of atomism derived from Lucretius and others, natural philosophers such as Pierre Gassendi, Robert Boyle, and Isaac Newton, among others, developed a mechanical philosophy of particles set in motion by a Christian God who created the world *ex nihilo*. A mathematical world described by geometry and calculus, in which equations could be solved and new situations predicted, provided truths of nature that could be verified by experiments.

The mathematical method is one pillar of modern science; experimentation the other. Together the two methods of mathematical derivation and empirical demonstration lead to an understanding of the natural world. It is that understanding that allows for the prediction of natural events and for the possibility of managing resources and inventing applications that improve life in the everyday world.

Pantheism

Spinoza's identification of God with Nature and *Natura naturans* with *Natura naturata* was a major influence on what came to be known as pantheism. While Spinoza may have been its greatest exponent, the English word "pantheist" was

114   *Controlling Nature*

coined by John Toland (1670–1722) in 1705. Toland, an Irishman, received his education at Glasgow and Edinburgh followed by Leiden and Oxford. In developing his philosophy of reverence for a materialistic cosmos, Toland was particularly influenced by Lucretius and Bruno. Like Spinoza, in his first book, *Christianity not Mysterious* (1696), he maintained that apparent mysteries and miracles in the *Bible* could be explained by natural causes.[40]

Toland engaged with Spinoza's philosophy in 1704 in his *Letters to Serena*, letters to Sophia Charlotte, Queen of Prussia. Two of these, "A Letter to a Gentleman in Holland," and "Motion Essential to Matter" were presented as an ostensible attack on Spinoza. Ian Leask argues, however, that Toland's apparent critique of Spinoza was actually a means of perfecting Spinozism by way of the philosophy of Leibniz. In 1705, he identified himself as a "Pantheist," in his "*Socinianism Truly Stated*, by a Pantheist." Without mentioning Spinoza's name, he wrote to Leibniz on February 14, 1710, about the "pantheistic opinion of those who believe in no other eternal being but the universe."[41]

Toland's foremost work on pantheism, published in 1720 two years before his death, was entitled *Pantheisticon, or the Form of Celebrating the Socratic Society* (*Pantheisticon, sive formula celebrandae sodalitatis socraticae*). Here he states that the universe is made of matter, and contains within it, its own principle of motion. We live in an infinite universe that has no center or limit, in which there are other worlds like the earth. Change is perpetual with all things in constant motion and development. The mind and soul are mortal and perish with the body. He famously responded to an innkeeper who asked him what country he was from, stating, "The Sun is my Father, the Earth my Mother, the World's my Country, and all Men are my Relations."[42]

The *Pantheisticon* was a work written in celebration of a "Socratic Society." In it he created a liturgy for meetings of this potentially secret group in which the President states, "All things in the world are One, and One is All in all things," to which the members respond, "What is All in all things is God, Eternal and Immense, Neither begotten or ever to perish."[43]

For Toland, as for Spinoza, the universe was one and the same as God, imbued with energy and reason. God was neither personal, nor transcendent, but immanent within nature as animate and alive. "[T]he Force and Energy of the Whole," he stated, "the Creator and Ruler of All, and always tending to the best End, is God, whom you may call the Mind, if you please, and Soul of the Universe." His secret society had a hymn which the members sang to the God in which we all live and move and who "animates all things, Forms, nourishes, increases, creates; Buries, and takes into itself all things."[44]

German philosopher Gotthold Ephraim Lessing (b. 1729–1781) was one of the first eighteenth-century writers to announce that Spinoza had greatly influenced his ideas about God and nature. In his dialogue with philosopher Friedrich Heinrich Jacobi, Lessing wrote that he took no pleasure in orthodox ideas of God, and in response to Jacobi's question, "[Y]ou were greatly at one with Spinoza?," he answered, "Did I rank myself with anyone, it were with none but him." The next day Jacobi queried, "I had no idea that I should find a Spinozist or a pantheist in

Natural Law   115

you, and still more that you should speak so unreservedly as you did." To which Lessing answered, "There is no philosophy but the philosophy of Spinoza."[45]

If Spinoza was the philosopher of pantheism and Toland its definer, Goethe is said to have been its poet. He wrote that "To discuss God apart from Nature is both difficult and perilous; it is as if we separated the soul from the body. We know the soul only through the medium of the body, and God only through Nature." He continued, "[E]verything which exists necessarily pertains to the essence of God, because God is the one Being whose existence includes all things."[46]

In 1869, T.H. Huxley, having been asked to write an introduction for the first issue of the scientific journal *Nature* (November 4, 1869) could think of nothing more appropriate than to quote Goethe:

> NATURE! We are surrounded and embraced by her: powerless to separate ourselves from her, and powerless to penetrate beyond her.
>
> Without asking, or warning, she snatches us up into her circling dance, and whirls us on until we are tired, and drop from her arms.
>
> She is ever shaping new forms: what is, has never yet been; what has been, comes not again. Everything is new, and yet nought but the old.
>
> We live in her midst and know her not. She is incessantly speaking to us, but betrays not her secret. We constantly act upon her, and yet have no power over her.

In a letter of 1828, Goethe states that he composed his essay on "Nature" sometime in 1786.

> I do not exactly remember having written these reflections, but they very well agree with the ideas which had at that time become developed in my mind. . . . There is an obvious inclination to a sort of Pantheism, to the conception of an unfathomable, unconditional, humorously self-contradictory Being, underlying the phenomena of Nature; and it may pass as a jest, with a bitter truth in it.[47]

Other Germans who are considered pantheists include philosophers Johann Gottlieb Fichte (1762–1814), Friedrich Shelling (1775–1854), and Georg Hegel (1770–1831) along with composer Ludwig von Beethoven (1770–1827). Americans include William Cullen Bryant (1794–1878), Ralph Waldo Emerson (1803–1882), Henry David Thoreau (1817–1862), Walt Whitman (1819–1892), Robinson Jeffers (1887–1962), and Ansel Adams (1902–1984). Scientists include Hans Christian Ørsted (1777–1851), Hermann von Helmholtz (1821–1894), Albert Einstein (1879–1955), and Isador Rabi (1898–1988). Spinoza's pantheistic philosophy has also been compared with the Vedanta tradition of Hinduism and other Eastern philosophies.[48]

Einstein was one of the most prominent followers of Spinoza. He believed in Spinoza's concept of *Deus sive Natura*, stating "We followers of Spinoza see our God in the wonderful order and lawfulness of all that exists." Einstein's God was

116   *Controlling Nature*

not a personal God, but one that was immanent within the natural world. In response to a question posed by Rabbi Herbert S. Goldstein in 1929 as to whether he believed in God, Einstein responded, "I believe in Spinoza's God who reveals himself in the orderly harmony of what exists, not in a God who concerns himself with fates and actions of human beings."[49]

Environmental Philosophy and Ethics

In the late twentieth century, Spinoza's philosophy became the foundation for the Deep Ecology movement. Its precepts were formulated by Norwegian philosopher Arne Naess (1912–2009) and promoted by California philosopher George Sessions who, like Naess, was a Spinoza scholar. They used Spinoza's *Deus sive Natura* as the basis for a new relationship between humans and nature. As Naess put it:

> The God of the *Ethics* may be identified essentially with Nature-as-creative (*natura naturans*)—the creative aspect of a supreme whole with two aspects, the creative and created—*natura naturata*. The latter are the existing beings in their capacity of being there, temporarily. There is creativity, but not a creator. The verb to nature (*naturare*) covers both aspects in its dynamic aspect. A comparable verb today would be "Gaia-ing," a term suitable for those who accept the most radical versions of the Gaia hypothesis: the planet Earth as a self-regulating, living being. Clearly such ideas are inspiring for radical environmentalists.[50]

For George Sessions, Spinoza's "knowledge of the laws of nature" gives humans a view of the world "under the aspect of eternity." The intellectual love of God goes beyond mathematical law and results in an intuitive knowledge of the laws of nature, "a rare mystical and non-communicable form of insight" that "gives direct knowledge of God or Substance." In grasping the laws of nature, to the extent to which we, as humans, are capable, we share in the "immortality and eternity of God/Nature/Substance." For the planet to survive, the dominant Western paradigm of mastery and domination must give way to a new Deep Ecology paradigm that embraces what Spinoza saw as "God/Nature in the sense of a vast organic divine Being."[51]

What was required, argued the two Spinozists, was a rethinking of Western ontology, epistemology, psychology, and ethics based on the idea of humans within rather than above nature. The Deep Ecology movement inspired numerous environmentalists around the world to act in ways consistent with the needs of nature and not just humans.[52]

Not everyone agreed, however, with the idea that Spinoza was an important inspiration for the current environmental movement. Spinoza scholar Genevieve Lloyd took issue with Naess and Sessions's arguments, stating that "it would be quite misplaced to claim Spinoza as patron philosopher of the environmental movement." Owing to Spinoza's views on animals, there was no evidence for an environmental ethic that would include the "rights of the non-human." In his

Ethic, Spinoza agreed with those who asserted (in contrast to Descartes) that animals feel pain, but argued that humans have "greater rights over beasts than beasts have over men." While animals have rights, as do humans, in the Hobbesian "state of Nature," this does not make them part of the human moral community as citizens of a state. Whereas the laws of nature pertain to the eternal order of nature, human reason and hence the moral law pertain only to humans and not to other animals. Lloyd concludes that the "theme of 'man as part of nature'" does not commit us to "extending the moral horizons to take account of the nonhuman."[53]

Naess responded that Spinoza did not use the terminology of morals, but rather a framework of joy, generosity, and sharing stemming from the human ability to reason. In Spinoza's system, humans and other animals did indeed have certain traits in common, such as procreation, feelings of pain, and the tendency to "persevere in its own essence." "If there is too little food, friends living according to reason quite naturally divide it among themselves, where rats try to get it all for themselves." Humans have the capability of coming to agreement through reason, whereas animals cannot. In conflicts of interest, however, humans will give themselves priority over animals. But, just as we are capable of acting toward other humans out of love and generosity, so we can act out of reason, generosity, fortitude, and love toward animals. When in Spinoza's framework, we act out of the intellectual love and understanding of God, we do so, in current terms, out of "knowledge of the intimate bonds between all living and non-living beings."[54]

Conclusion

Baruch Spinoza's rationalism was a harbinger of the Enlightenment's focus on reason. Everything in the natural world had a rational and not a mystical explanation. The *Bible* contained no miracles, nor did God perform any miracles. God and Nature were, in fact, one and the same, identified as *Deus sive Natura.* Together they constituted Substance, or reality. God was immanent within the cosmos, not transcendent above it. Logic, reason, and mathematics were the keys to understanding the universe. *Natura naturans* was a single eternal, self-causing substance that was one with God; *Natura naturata* was the created world described by mathematics. Nowhere in Spinoza's philosophy was there a place for chaotic, unpredictable events. The idea that God might create earthquakes, volcanic eruptions, plagues, droughts, and storms in retribution for human sin was simply a projection onto the world, instigated by human irrationality and emotional frailty.

Spinoza's rational philosophy led to the Enlightenment idea expressed most cogently by Pierre-Simon Laplace in his alleged answer to Napoleon's question about the role of God in the regulation of the heavens, to which he responded, "Sire, I had no need of that hypothesis."[55] Spinoza's *Deus sive Natura* stimulated the development of pantheism. His logical method of argument that drew on Euclidean theorems, propositions, and axioms reflected the evolving importance played by mathematics in the systems of Gottfried Wilhelm Leibniz, Isaac Newton, and other natural philosophers.

118   *Controlling Nature*

Notes

1 Matthew Stewart, *The Courtier and the Heretic: Leibniz, Spinoza and the Fate of God in the Modern World* (New York: Norton, 2006), pp. 29–30: Some of Spinoza's peers said that

> he believed that the books of Moses were made by man; that the soul dies with the body; and that God is a corporeal mass. For Jews of the time, just as much as for Christians, such notions were frightening heresies. . . . In his mature works, Spinoza does in fact suggest that the *Bible* is a human invention, in a manner of speaking; and he explicitly rejects the doctrine of personal immortality. While he nowhere says that God is a part of the corporeal world, he does indeed claim that the corporeal world is a part of God (to put it crudely). (pp. 29–30.)

On Spinoza's life see, Margaret Gullan-Whur, *Within Reason: A Life of Spinoza* (London: Jonathan Cape, 1998), esp. pp. 63–71; Steven Nadler, *Spinoza: A Life* (Cambridge, UK: Cambridge University Press, 1999), esp. pp. 134–137, 153 and illustrations following p. 146 showing the "Jewish Quarter in Amsterdam," the "Amsterdam Portuguese Synagogue (interior, 1639–75)," and the "Amsterdam Portuguese Synagogue (exterior), 1639–75." Also H.G. Hubbeling, "Spinoza's Life: A Synopsis of the Sources and Some Documents," *Giornale Critico della Filosofia Italiana* 8, no. 3–4 (1977): 390–409; Jean Colerus, *La Vie de B. de* Spinoza (1706), pp. 7–91 and [Jean Maximilien] Lucas, *La Vie de M. Benoit de Spinoza* (1719, 1735), pp. 93–132, in *Vies de Spinoza* (Paris: Allia, 1999). On Jean Maximilien Lucas, see also, "The Life of the Late M. de Spinoza," in *The Oldest Biography of Spinoza*, ed. and trans. A[braham] Wolf (London: Allen & Unwin, 1927), pp. 39–75, esp. pp. 44–51 on Spinoza's appearance before the judges, the accusations against him, and his excommunication. When his so-called friends (pp. 47–48):

> saw the proper time to push it [the allegations] more actively they made their report to the Judges of the Synagogue, whom they incited in such a manner that they thought of condemning him without hearing him first. When the ardour of the first flare had passed (the holy ministers of the Temple are not exempt from wrath), they had him summoned to appear before them. On his part, feeling that his conscience had nothing to reproach him, he went cheerfully to the Synagogue.

On the allegation that Spinoza believed that God has a body, see p. 45: "I confess, said the disciple [Spinoza] since nothing is to be found in the Bible about the non-material or incorporeal there is nothing objectionable in believing that God is a body."

2 Stewart, *The Courtier and the Heretic*, pp. 30–35, quotation on p. 33.

3 Stewart, *The Courtier and the Heretic*, pp. 54–61.

4 Stewart, *The Courtier and the Heretic*, pp. 18–28.

5 Stewart, *The Courtier and the Heretic*, pp. 28–32, 60.

6 Benedictus de Spinoza, *Spinoza Opera*, ed. Karl Gebhardt, 4 vols. (Heidelberg: C. Winters, [1925]); Baruch Spinoza, *Principles of Cartesian Philosophy*, trans. Harry E. We (New York: Philosophical Library, 1961); Spinoza, *The Principles of Cartesian Philosophy and Metaphysical Thoughts*; Followed by Inaugural dissertation on matter; Lodewijk Meyer; trans. by Samuel Shirley, introduction and notes by Steven Barbone and Lee Rice (Indianapolis, IN: Hackett, 1998); Spinoza, *Theological-Political Treatise*, ed. Jonathan Israel, trans. Michael Silverthorne and Jonathan Israel (New York: Cambridge University Press, 2007); Spinoza, *Ethics* (1677), ed. and trans. Edwin Curley (New York: Penguin, 1996).

7 Stewart, *The Courtier and the Heretic*, pp. 126–127.

8 Benedict de Spinoza, *A Theologico-Political Treatise and A Political Treatise*, trans. R.H.M. Elwes (London: George Bell, 1900 [1883]; repr., New York: Dover, 1951; Cosimo, 2007); see Ch. 5 and 16, quotations on pp. 210, 200, 73, 74, 206; Jari Niemi, "Spinoza's Political Philosophy," *Internet Encyclopedia of Philosophy*, http://www.iep.utm.edu/spin-pol/.

Natural Law   119

9 Baruch Spinoza, *Ethic*, in *Selections*, ed. John Wild (New York: Scribner's, 1930), Part IV, Proposition 73, quotations on pp. 351, 352.

10 George Sidney Brett, *The Philosophy of Gassendi* (London: MacMillan, 1908), p. 19; Pierre Gassendi, *Syntagma philosophicum* in *Opera Omnia*, 6 vols. (Stuttgart-Bad Cannstatt: Froman, 1954 [1658]), facsimile edition; Pierre Gassendi, *The Selected Works of Gassendi*, trans. Craig B. Bush. London: Johnson Reprint, 1972.

11 Baruch Spinoza, *Principles of Descartes's Philosophy*, trans. Halbert Hains Britan (LaSalle, IL: Open Court, 1961 [1663]; repr., 2010). See also James Iverach, *Descartes, Spinoza, and the New Philosophy* (New York: Scribner's, 1904).

12 René Descartes, *Discourse on Method* (1636), in *The Philosophical Works of Descartes*, ed. E. S. Haldane and G.R.T. Ross, 2 vols. (New York: Dover, 1955), vol. 1, p. 85; see also Descartes, *Meditations on First Philosophy* (1643), ibid., vol. 1, pp. 131–199.

13 René Descartes, *Le Monde, ou Traite de la lumiere*, trans. with an introduction by Michael Sean Mahoney (New York: Abaris Books, 1979; portions published 1644 in the *Principia philosophiae* [*Principles of Philosophy*] and in toto, 1677), see Ch. 6 (p. 55); Descartes further described the origins of chaos and its meaning:

> But, before I explain this at greater length, stop again for a bit to consider that chaos, and note that it contains nothing that is not so perfectly known to you that you could not even pretend not to know it. For, as regards the qualities that I have posited there, I have, if you have noticed, supposed them to be only such as you can imagine them. And, as regards the matter from which I have composed the chaos, there is nothing simpler nor easier to know among inanimate creatures. The idea of that matter is so included in all those that our imagination can form that you must necessarily conceive of it or you can never imagine anything. (p.55)

> For another translation see, René Descartes, *The World and Other Writings*, ed. Stephen Gaukroger (Cambridge, UK: Cambridge University Press, 1998), "The Treatise on Light," Ch. 6, "Description of a new world, and the qualities of the matter of which it is composed," [Gaukroger's note: in the 1664 edition, "Description of a New World, very easy to know, but nevertheless similar to ours, and even to the chaos which the poets imagined to have preceded it"], pp. 23–24:

> For God has established these laws in such a marvelous way that even if we suppose that He creates nothing more than what I have said, and even if He does not impose any order or proportion on it but makes it of the most confused and muddled chaos that any of the poets could describe, the laws of nature are sufficient to cause the parts of this chaos to disentangle themselves and arrange themselves in such a good order that they will have the form of a most perfect world, a world in which one will be able to see not only light, but all the other things as well, both general and particular, that appear in the actual world. But before I explain this at greater length, pause again for a minute to consider this chaos, and note that it contains nothing which you do not know so perfectly that you could not even pretend to be ignorant of it. For the qualities that I have placed in it are only such as you could imagine. And as far as the matter from which I have composed it is concerned, there is nothing simpler or more easily grasped in inanimate creatures. The idea of that matter is such a part of all the ideas that our imagination can form that you must necessarily conceive of it, or you can never imagine anything at all.

14 Descartes, *Le Monde*, trans. Mahoney, Ch. 7, p.59. See also Descartes, *The World and Other Writings*, ed. Gaukroger, Ch. 7, "The Laws of Nature of this new world," pp. 24–25:

> But I do not want to delay any longer telling you the means by which Nature alone is able to untangle the confusion of the chaos which I have been speaking about, and what the Laws of Nature that God has imposed on it are. Take it then, first, that by "Nature" here I do not mean some deity or other sort of imaginary power. Rather,

120 *Controlling Nature*

> I use the word to signify matter itself, in so far as I am considering it taken together with the totality of qualities I have attributed to it, and on the condition that God continues to preserve it in the same way that He created it.

15 Spinoza, *Principles of Descartes's Philosophy*, pp. 40–41, 41–44, 53, 57 ff. In Meditation II of his *Meditations on First Philosophy* (op. cit., Haldane and Ross, ed., vol. 1, p. 157), Descartes had distinguished between the faculties of reason and the imagination, arguing that

> bodies are not properly speaking known by the senses or by the faculty of imagination, but by the understanding only, and since they are not known from the fact that they are seen or touched, but only because they are understood, I see clearly that there is nothing which is easier for to know than my mind (p.157).

On the limitations of the senses and the imagination and the significance of reason, see also Descartes, *Discourse on Method*, ibid., Haldane and Ross, ed., Part II, pp. 91–94 and Part IV, pp. 101–105, esp. p. 104; *Passions of the Soul*, trans. and annotated by Stephen Voss (Indianapolis, IN: Hackett, 1989), Part I, Articles 19–20, pp. 340–341.

16 Spinoza, *The Cogitata Metaphysica*, Appendix to Spinoza, *Principles of Descartes's Philosophy*, see pp. 160, 164.

17 Stewart, *The Courtier and the Heretic*, pp. 100–102.

18 Spinoza, *Theological-Political Treatise*, trans. Elwes, pp. 81, 82. See also Spinoza, *Theological-Political Treatise*, ed. Israel, trans. Silverthorne and Israel, Ch. 6, "On Miracles," sec. 1, quotations on pp. 81–82: "[T]hey call unusual works of nature miracles or works of God and do not want to know the natural causes of things. . . ." They "ascribe all things to his will and governance, by ignoring natural causes and evincing wonder at what is outside the normal course of nature and revere the power of God best when they envisage the power of nature as if it were subdued by God."

19 Spinoza, *Theological-Political Treatise*, trans. Elwes, quotation on pp. 83–84. See also Spinoza, *Theological-Political Treatise*, ed. Israel, trans. Silverthorne and Israel, Ch. 6, "On Miracles," see sec. 7, p. 85, sec. 12, p. 89, sec. 14, p. 90; quotations from p. 84: "For otherwise what we are saying is that God has created a nature so impotent and with laws and rules so feeble that He must continually give it a helping hand."

20 Spinoza, *Theological-Political Treatise*, trans. Elwes, quotation on p. 92. See also Spinoza, *Theological-Political Treatise*, ed. Israel, trans. Silverthorne and Israel, Ch. 6, "On Miracles," secs. 14, 15, quotations from sec. 15, p. 91: "We can conclude without reservation that all things that are truly reported to have happened in Scripture necessarily happened according to the laws of nature. . . ." If anything can be found that does contradict the laws of nature, then "we must accept in every case that it was interpolated into the Bible by blasphemous persons. For what is contrary to nature, is contrary to reason, and what is contrary to reason, is absurd, and accordingly to be rejected."

21 Richard H. Popkin, "Hume and Spinoza," *Hume Studies* 5, no. 2 (November 1979): 65–93, see esp. pp. 76 and 82, quotation on p. 76.

22 Stewart, *The Courtier and the Heretic*, pp. 104–105.

23 Henry A. Lucks, "*Natura Naturans—Natura Naturata*," *The New Scholasticism* 9, no. 1 (1935): 1–24; John McPeek, *Bruno, Spinoza, and the Terms* Natura Naturans *and* Natura Naturata (New Orleans: Tulane University, 1965); Thomas Aquinas, *Summa Theologica*, Latin text and English trans. (London: Blackfriars, 1963 [1265–1274]), vol. 26, IaIIae, 85, 6, reply, p. 101 (Latin, p. 100, note 9).

24 McPeek, *Bruno, Spinoza, and the Terms* Natura Naturans *and* Natura Naturata, pp. 38–44.

25 Lucks, "*Natura Naturans—Natura Naturata*," see pp. 10 and 21–24 on Spinoza's possible scholastic sources. See also McPeek, *Bruno, Spinoza, and the Terms* Natura Naturans *and* Natura Naturata, pp. 4–5, 35; McPeek cites the following references as to Spinoza's knowledge of scholasticism and Bruno: Frederick Pollock, *Spinoza*

Natural Law 121

(London: Duckworth, 1935), pp. 12, 49; B[enjamin] A[pthorp] G[ould] Fuller, *A History of Philosophy*, ed. Sterling M. McMurrin (New York: Holt, 1960 [1955]), vol. 2, p. 71; Rudolf Kayser, *Spinoza: Portrait of a Spiritual Hero* (New York: Philosophical Library, 1946, p. 105; Harald Høffding, *A History of Modern Philosophy*, trans. B. E. Meyer (London: MacMillan, 1924), vol. 1, p. 296; John Caird, *Spinoza* (Edinburgh: Blackwood, 1888), pp. 89–90; Constance E. Plumptre, *General Sketch of the History of Pantheism* (London: Beacon, 1878 and 1879), vol. 1, p. 355; Dorothea Waley Singer, *Giordano Bruno: His Life and Thought, with Annotated Translation of His Work*, On the Infinite Universe and Worlds (New York: Schuman, 1950). See also John Deely, *Four Ages of Understanding: The First Postmodern Survey of Philosophy from Ancient Times to the Turn of the Twenty-First Century* (Toronto: University of Toronto Press, 2001), pp. 135–140. Deely (p. 138) in his discussion of John Scotus Erigena and the Latin translators of Averroes's *Commentary on Aristotle* notes the oblique influence of Erigena's fourfold division of nature on the twofold division of *natura naturans* and *natura naturata* in the Latin world around the turn of the thirteenth century.

26 Fuller, *History of Philosophy*, vol. 2, p. 33, quoted in McPeek, p. 21; McPeek, pp. 10, 20, 29–31, 36–37.

27 Baruch Spinoza, *Spinoza's Short Treatise on God, Man, and His Well-Being*, trans. and ed. A[braham] Wolf (London: Adam and Charles Black, 1910), Ch. 8, p. 56 (quotation) and note p. 199–200; see also Spinoza, "Short Treatise on God, Man and His Well-Being," in *Selections*, ed. Wild, p. 80; Yitzhak Melamed, "Spinoza's Metaphysics of Substance: The Substance-Mode Relation as a Relation of Inherence and Predication," *Philosophy and Phenomenological Research* 78, no. 1 (January 2009): 17–82. Melamed concludes (p. 77): "Our close examination of a significant body of texts show that Spinoza considered particular things, such as Mt. Rushmore and Napoleon, to be modes inhering in God, and that Spinoza was a pantheist."

28 Spinoza, *Spinoza's Short Treatise*, ed. Wolf, Ch. 9, p. 57.

29 Spinoza, *Cogitata Metaphysica*, Appendix to idem, *Principles of Descartes's Philosophy*, pp. 160, 164.

30 Lucia Lermond, *The Form of Man: Human Essence in Spinoza's* Ethic (Leiden: E. J. Brill, 1988), Ch. 2, "*Natura Naturans* and *Natura Naturata*," p. 6; Lucks, "*Natura Naturans—Natura Naturata*," see p. 10. Spinoza, *Ethic*, in *Selections*, ed. Wild, Part I, Proposition 29, Scholium, p. 126; see also, Spinoza, *Ethics* (1677), ed. and trans. Edwin Curley (New York: Penguin, 1996), Part I, "Of God," Propositions 29 and 31 and Iverach, *Descartes, Spinoza, and the New Philosophy*, pp. 190, 197–199.

31 Spinoza, *Ethic*, in *Selections*, ed. Wild, Part I, Proposition 29, Scholium, p. 126; for the Latin see Spinoza, *Spinoza Opera*, ed. Gebhardt, vol. 2, p. 71, line 5:

> Antequam ulterius pergam, hic, quid nobis per Naturam naturantem, & quid per Naturam naturatam intelligendum sit, explicare volo, vel potius monere. Nam ex antecedentibus jam constare existimo, nempe, quod per Naturam naturantem nobis intelligendum est id, quod in se est, & per se concipitur, sive talia substantiae attributa, quae aeternam, & infinitam essentiam exprimunt, hoc est. . . ., Deus, quatenus, ut causa libera, consideratur. Per naturatam autem intelligo id omne, quod ex necessitate Dei naturae, sive uniuscujusque Dei attributorum sequitur, hoc est, omnes Dei attributorum modos, quatenus considerantur, ut res, quae in Deo sunt, & quae sine Deo nec esse, nec concipi possunt.

32 Alfred Weber, *History of Philosophy*, trans. Frank Thilly (New York: Charles Scribner's Sons, 1896), sec. 55, pp. 323–343, quotations on pp. 329, 327–328, 332, 327.

33 Spinoza, "Letter 12, to Ludovicus Meyer (On the Nature of the Infinite)," in Spinoza, *Selections*, ed. Wild, pp. 410–417, quotation on p. 412. On modes see Spinoza, *Ethic*, *Selections*, ed. Wild, Part I, Proposition 22, 23, pp. 121–122. Hermann De Dijn, "The Articulation of Nature, or the Relation God-Modes in Spinoza," *Giornale Critico della Filosofia Italiana* 8, no. 3–4 (1977): 337–344:

122 *Controlling Nature*

> The problem of the relation between *Natura Naturans* and *Natura Naturata*, or the problem of the relation between God and his modes or effects, can be expressed as the problem of the relation between God's immanence and his transcendence vis à vis his modes. . . . As the substance constituted by an infinity of attributes, as *Natura Naturans*, God *necessarily* produces an infinity of things in an infinity of ways (*Ethic*, I, Prop. 16), i.e. an infinity of modes in each of the infinity of attributes (the *Natura Naturata*). (p. 337) Nevertheless, this does not mean that a mode is God himself (*Deus*), it is God only *in so far as* he is modified in one or other of his attributes (*Deus quatenus*). The fact that a mode in a certain sense can be said to be substance, does not allow us to equate it with or [as] substance: it is only *Deus quatenus*. (p. 338)

34 Spinoza, "Letter 12," in *Selections*, ed. Wild, quotations on p. 412.
35 Donald Rutherford, "Spinoza's Conception of Law," in *Spinoza's Theological–Political Treatise: A Critical Guide*, ed. Yitzhak Y. Melamed and Michael A. Rosenthal (New York: Cambridge University Press, 2010), Ch. 8, pp. 143–167, see pp. 143–144; Spinoza, *Theological-Political Treatise*, ed. Israel, trans. Silverthorne and Israel, Ch. 4, "On the Divine Law," p. 57, sec. 1, quotations from lines 52–53, 56–61; see also Jon Miller, "Spinoza and the Concept of a Law of Nature," *History of Philosophy Quarterly* 20, no. 3 (July 2003): 257–276.
36 Rutherford, "Spinoza's Conception of Law," pp. 144–145; Spinoza, *Theological-Political Treatise*, trans. Silverthorne, ed. Israel, Ch. 4, "On the Divine Law," sec. 1, p. 57, quotations from lines 54–55 and 62–64.
37 Stewart, *The Courtier and the Heretic*, pp. 100–102.
38 Spinoza, "Letter 12," to Ludovicus Meyer, in *Selections*, ed. Wild, pp. 413–414. Spinoza's forms of knowledge were ranked as the rational, the intuitional, and the mathematical. See Kevin Von Duuglas-Ittu, http://kvond.wordpress.com/2009/09/24/by-mathematical-attestation-spinozas-use-of-mathematics/.
39 Spinoza, "Letter 12," in *Selections*, ed. Wild, pp. 413–414, quotation on p. 413.
40 Rosalie L. Cole, "Spinoza and the Early English Deists," *Journal of the History of Ideas* 20, no. 1 (January 1959): 23–46, writes, "Toland appears to have coined the word 'pantheist' in 1705; so far as I know, he himself never used the noun 'pantheism' in his writings" (p. 45, note 88). English mathematician Joseph Raphson used the term "pantheismus" in his *De Spatio Reali* (1697), see http://pantheist.weebly.com/toland.html and Ann Thomson, *Bodies of Thought: Science, Religion, and the Soul in the Early Enlightenment* (New York: Oxford University Press, 2008), p. 54. John Toland, *Christianity not Mysterious or, A treatise shewing that there is nothing in the Gospel contrary to reason, nor above it: and that no Christian doctrine can be properly call'd a mystery* (London: Printed for Sam Buckley, 1696); on Toland and other pantheists, see Margaret C. Jacob, *The Radical Enlightenment: Pantheists, Freemasons, and Republicans* (London: Allen & Unwin, 1981), Ch. 7, pp. 215–255.
41 John Toland, *Letters to Serena: containing, I. The origin and force of prejudices. II. The history of the soul's immortality among the heathens. III. The origin of idolatry, and reasons of heathenism, As also, IV. A letter to a gentleman in Holland, showing Spinosa's system of philosophy to be without any principle or foundation. V. Motion essential to matter; an answer to some remarks by a nobel friend on the confutation of Spinosa. To all which is prefixed, VI. a preface; being a Letter to a gentleman in London* (London: Printed for M. Cooper; W. Reeve; and C. Sympson, 1753), see Letter IV, p. 131, Letter V, p. 163; Ian Leask, "Unholy Force: Toland's Leibnizian 'Consummation' of Spinozism," *British Journal for the History of Philosophy* 20, no. 3 (2012): 499–537. John Toland, *A Collection of Several Pieces of Mr. John Toland, Now first published from his Original Manuscripts, With Some Memoirs of his Life and Writings*, 2 vols. (London: Printed for J. Peale, 1726), vol. 2, "Letter to Leibniz," February 14, 1710, p. 388–394, see p. 394.

Natural Law 123

42 John Toland, *Pantheisticon, sive formula celebrandae sodalitatis socraticae* (London, 1720); *Pantheisticon, or, the form of celebrating the Socratic-Society. Divided into three parts. . . . To which is prefix'd a discourse upon the antient and modern societies of the learned, . . . And subjoined, a short dissertation upon a two-fold philosophy of the pantheists, . . . Written originally in Latin, by the ingenious Mr. John Toland. And now, for the first time, faithfully rendered into English* (London: Printed for Sam. Paterson; and sold by M. Cooper, 1751), pp. 15, 17, note, quotation on p. 33.

43 Toland, *Pantheisticon*, p. 63, quotations on p. 70.

44 Toland, *Pantheisticon*, pp. 17, 71, quotation on 71.

45 Constance E. Plumptre, *General Sketch of the History of Pantheism*, 2 vols. (London: Beacon, 1878–79), vol. 2, pp. 104–105; Gerard Vallee, trans., *The Spinoza Conversations Between Lessing and Jacobi, Text with Excerpts from the Ensuing Controversy*, introduction by Gerard Vallee, trans. G. Vallee, J. B. Lawson, and C. G. Chapple (Lanham, MD: University Press of America, 1988). The two-volume work by a nineteenth-century woman scholar, Constance E. Plumptre, was originally published anonymously in 1878–79 and then under the pseudonym Charles E. Plumptre. See http://orlando.cambridge.org/public/svPeople?person_id=plumce. C. E. Plumptre wrote during the later nineteenth century on religious and philosophical issues. She authored five books and many periodical essays. She was especially interested in pantheist philosophers, deliberately seeking out and championing persecuted and obscure thinkers of the past such as Giordano Bruno and Lucilio Vanini. In 1878, C.E.P. anonymously published the first volume in a two-volume series of philosophical thoughts: *General Sketch of the History of Pantheism*. She died ca. March 12, 1929, at just under eighty years of age.

46 Plumptre, *History of Pantheism*, vol. 2, p. 261, quoting from George Henry Lewes, *Life of Goethe*, 2 vols. (New York: F. Ungar, 1864), pp. 71–73.

47 T. H. Huxley, "Goethe: Aphorisms on Nature," first issue of *Nature*, November 4, 1869, including letter of Goethe to Chancellor von Muller, May 26, 1828. http://www.nature.com/nature/about/first/aphorisms.html.

48 For a list of pantheists, see http://en.wikipedia.org/wiki/List_of_pantheists, and Paul Harrison, "Theological Notes: A Promising Time for Pantheism," *The Independent*, May 21, 1999, http://www.independent.co.uk/news/people/theological-notes-a-promising-time-for-pantheism-1094869.html; on pantheism and Eastern philosophies, see John Hunt, *An Essay on Pantheism* (London: Longmans, Green, Reader, and Dyer, 1866) and Plumptre, *History of Pantheism*, vol. 1, Ch. 1–5.

49 On Spinoza's influence on Einstein, see http://en.wikipedia.org/wiki/Baruch_Spinoza, and Roger Scruton, *Spinoza: A Very Short Introduction*, 1986 (Oxford, UK: Oxford University Press, 2002).

50 Arne Naess, *Spinoza and the Deep Ecology Movement* (Delft, Netherlands: Eburon, 1993), reprinted in Harold Glasser and Alan Drengson, eds., *The Selected Works of Arne Naess*, 10 vols. (Dordrecht, Netherlands: Springer, 2005), vol. 10, pp. 395–420, quotation on pp. 402–403; (reprinted in Arne Naess, *Ecology of Wisdom: Writings by Arne Naess*, ed. Alan Drengson and Bill Devall (Berkeley, CA: Counterpoint Press, 2008), pp. 230–251, quotation on p. 237); Arne Naess, "Spinoza and Ecology," *Philosophia* 7 (1977): 45–54. Other environmental philosophers who have been influenced by Spinoza include Eccy De Jonge, *Spinoza and Deep Ecology: Challenging Traditional Approaches to Environmentalism* (Burlington, VT: Ashgate, 2004); Freya Mathews, *The Ecological Self* (London: Routledge, 1991); Bill Devall, *Simple in Ends, Rich in Means: Practicing Deep Ecology* (Salt Lake City, UT: Peregrine Smith Books, 1988).

51 George Sessions, "Spinoza and Jeffers on Man and Nature," *Inquiry* 20 (1977): 481–528, quotations on p. 501, 502, 507.

52 Arne Naess, "The Shallow and the Deep, Long-Range Ecology Movement: A Summary," *Inquiry* 16, no. 1 (1973): 95–100; Bill Devall and George Sessions, eds., *Deep Ecology: Living as if Nature Mattered* (Salt Lake City, UT: G. M. Smith, 1985).

124 *Controlling Nature*

53 Genevieve Lloyd, "Spinoza's Environmental Ethics," *Inquiry* 23, no. 3 (1980): 293–311, quotations on pp. 294; p. 295, quoted from Spinoza, *Ethic*, Part 4, Proposition 37 (note 1); p. 297; p. 298 (from Spinoza's *Theological-Political Treatise*, ed. Israel, trans. Silverthorne and Israel, vol. I, p. 202); p. 309. See also Lloyd, *Part of Nature: Self-Knowledge in Spinoza's* Ethics (Ithaca, NY: Cornell University Press, 1994).

54 Arne Naess, "Environmental Ethics and Spinoza's *Ethics*: Comments on Genevieve Lloyd's Article," *Inquiry* 23, no. 3 (1980): 313–325, see pp. 317, 318, 319, 323, quotations on 319, 323.

55 On Laplace's alleged quotation, "Sire, I had no need of that hypothesis," see "Pierre-Simon LaPlace" Wikipedia, (https://en.wikipedia.org/wiki/Pierre-Simon_Laplace). It is alleged that LaPlace actually made the remark in reference to whether God (as Newton believed) intervened in the Universe to keep the system working properly.

6 Laws of Nature
Leibniz and Newton

In November 1676, three months before Spinoza's death, he was visited by the young Gottfried Wilhelm Leibniz. Leibniz, who was thirty years old at the time, arrived in The Hague for the specific purpose of meeting Spinoza. They spent three days together in intense conversation. According to philosopher Matthew Stewart, "the meeting with Spinoza was the defining event of Leibniz's life." Although Leibniz would later publically disavow that influence, the outcome was the recognition that a rational universe governed by the laws of nature held the keys to understanding the natural world.[1]

Stewart depicts Leibniz's arrival as follows:

> On or around November 18, 1676 . . . the thirty-year-old inventor of the calculus, the former privy-counselor of Mainz, and the newly appointed librarian to the Duke of Hanover stepped ashore, arms flapping, wig billowing, perfume dissipating in the autumn wind, and gamboled in his awkward way along the leaf-strewn canals toward the door of the house where Spinoza lived.[2]

There, at Spinoza's shared home, the two conversed "many times and at great length" in the course of three or more days. Leibniz's encounter with Spinoza in 1676 had repercussions that lasted for the rest of his life. During the years following their conversations in The Hague, Leibniz sought to differentiate himself and his philosophical system from Spinozism.[3]

In this chapter I argue that mathematical predictability in the philosophies of Gottfried Wilhelm Leibniz (1646–1716) and Isaac Newton (1642–1727) seemed to offer an antidote to the problem of chaos and unruly nature. I compare and contrast the philosophies of Spinoza, Leibniz, and Newton as to the roles of nature, activity, and God in the search for a rational, predictable world. Although neither Leibniz or Newton used the terms *natura naturans* or *natura naturata*, both described the created world (*natura naturata*) in terms of mathematics and both were deeply engaged with underlying sources of activity associated with *natura naturans*.

Figure 6.1 Spinoza's House in The Hague Where He Lived in 1676.

The House (marked with an x) in The Hague, owned by the painter Hendrik Van Der Spyck, in which Leibniz visited Spinoza in 1676. Spinoza died there on February 21, 1677, of tuberculosis resulting from his glass-grinding profession.
From Jean Maximillien Lucas, *The Oldest Biography of Spinoza*, trans. A[braham] Wolf (London: George Allen & Unwin, 1927), facing p. 72.

Rationalism

By the late seventeenth century, Spinoza, Leibniz, Newton, and others had established rationalism as the public face of natural philosophy. Mathematics and metaphysics were keys to the differing forms that rationalism took. Whereas Spinoza had viewed God as impersonal and immanent within nature, for Leibniz and Newton, God was personal and transcendent above it. Leibniz's rational God showed the primacy of his intellect through the perfection of his initial creation. Newton's God showed his will and care by his rational intervention in the created world. While none of the three allowed a chaotic nature to taint their mathematical analyses of the created world, all included principles of activity within their metaphysics. Spinoza did so through a monistic pantheism that characterized the self-acting world, Leibniz through his self-acting monads underlying the phenomenal world, and Newton through his active principles that continually reinvigorated a world prone to decline and decay.

Newton and Leibniz's mathematical descriptions of nature applied best to closed, mechanical systems operating under idealized conditions. Here

Figure 6.2 Gottfried Wilhelm Leibniz (1646–1716).
Copper engraving, 1775, by Johann Friedrich Bause (1738–1814), after a painting 1703, by Andreas Scheits (ca. 1655–1735)
Courtesy of akg-images.

prediction and control achieved enormous success in succeeding centuries. The limitations of mechanism were most blatant, however, in the prediction of irregularly occurring events such as earthquakes, volcanic eruptions, and plagues. These sudden, chaotic acts of nature *seemed* to defy mechanism's rational system of prediction and control. At the end of the chapter, I illustrate mechanism's explanatory limits by the massive earthquake that struck Lisbon, Portugal, in 1755.[4]

Figure 6.3 Isaac Newton (1642–1727) in 1689 (age 46).
Portrait by Godfrey Kneller, 1689.
Courtesy of the trustees of the Portsmouth Estates, UK and Jeremy Whitaker.

Leibniz and Spinoza

By the time of Leibniz's visit to The Hague in November 1676, Spinoza had published his anonymous *Tractatus Theologico-Politicus* (1670) and completed his *Ethics*, published posthumously in 1677. Leibniz had discovered the calculus and introduced the symbols ∫ and *dx*, invented his calculating machine comprising a box with gears and dials that would add, subtract, multiply, and divide, and had published a major mathematical work on the *Art of Combinations* (1666). He had studied with Christiaan Huygens in Paris where he met Nicholas Malebranche and Antoine Arnauld and at the age of twenty-nine had been admitted to the French Academy of Sciences. And he had read Spinoza's anonymously published *Tractatus*.[5]

Leibniz's interest in meeting Spinoza was sparked by his friendship with Walther Ehrenfried von Tschirnhaus, who had studied Spinoza's published work and was introduced to him via a letter from Henry Oldenburg in London. Tschirnhaus rekindled Leibniz's interest in Spinoza's *Tractatus* and its argument that God and Nature were one and the same. Tschirnhaus was also in possession of a copy of Spinoza's unpublished manuscript on *Ethics*, entrusted to him by Spinoza, which Leibniz was anxious to read and for which he sought permission from Spinoza via Tschirnhaus. Spinoza declined to allow Leibniz access to the manuscript, but Tschirnhaus nevertheless divulged many of its ideas in conversations with Leibniz. In November 1676 after a visit to Oldenburg in London and bearing a letter of introduction from Oldenburg, Leibniz traveled to Holland where he met several of Spinoza's contacts who smoothed the way for the visit to Spinoza in The Hague.[6]

In 1678, soon after Spinoza's *Opera Posthuma* had been published, Leibniz began to catalogue the assumptions with which he disagreed. Those included the ideas that God is a substance of which all creatures are modes, that God has no will, and that there is no such thing as life after death. All these ideas were, in the dominant theologies of the time, blatant heresies. In marginal notes made on reading Spinoza's *Opera Postuma*, Leibniz notes time and again that this or that proposition is fallacious, obscure, questionable, pretentious, or empty.[7]

The main point of difference involved the nature of God. Spinoza's God was one and the same as Nature, a nonanthropocentric, infinite, strictly rational God without personal characteristics—a God with whom one could not converse and to whom one could not pray. The world itself was neither good nor evil. The existing world followed from God's own nature. Leibniz's God, although equally rational, was, by contrast, a person. Because God is rational, "He" has a choice of which world to create. He chooses that world which is not only the most perfect, but is, in fact, the best of all possible worlds. It is a world that has both good and evil within it. Spinoza's God has no choice; Leibniz's God does. Spinoza's God is immanent; Leibniz's God is transcendent.[8]

Several months before his meeting with Spinoza, on February 11, 1676, Leibniz had already defined his ideas about God in clear opposition to Spinoza's *Tractatus*:

> God is not a kind of imaginary metaphysical being, incapable of thought, will, and action, as some make him. This would be the same as to say that God is nature, fate, fortune, necessity, or the world. But God is a definite substance, a person, a mind.[9]

His "Dissertation on the Art of Combinations" (1666), a publication that resulted from his doctoral dissertation, had begun a lifelong goal of defining a logical system of symbols and calculations that could be applied to all fields of thought. Such a system would introduce a rational method by which all debates could be resolved. Tschirnhaus's friendship inspired Leibniz to pursue his passion for mathematics.[10]

Sometime during the fall of 1675, after the arrival of Tschirnhaus, Leibniz put together the elements of the calculus including the use of the symbols $\int$ and dx to stand for the integral and derivative. He wrote to Henry Oldenberg on December 28, 1675, that

130   *Controlling Nature*

insofar as nature itself, to the degree to which it is known and can be sub-jected to this calculus and to the degree that new qualities are discovered and reduced to this mechanism, will also give to geometricians new material to which to apply it.[11]

Leibniz's concern with the mathematical description of nature centered around what he called the labyrinth of the continuum. In February 1676, he reflected on the problem of the continuum that would become the foundation for his math-ematics and metaphysics:

> The whole labyrinth about the composition of the continuum must be unraveled as rigorously as possible. . . . We must see whether it can be demon-strated that there is something infinitely small yet not indivisible.[12]

Just as points could not be added to each other to create a line, so material points could not be added together in ways that would account for the activity within the material world. Something immaterial, a substance, was required to explain activity. But both the activity of observed phenomena and the immateri-ality of the mental needed a unified explanation. What would ultimately emerge from this "labyrinth of the continuum" were the fundamentals of his philosophy. Substance comprised an infinity of self-acting monads whose unfolding lives existed in preestablished harmony with the lives of all the other monads and a phenomenal world that was well-founded (*phenomena bene-fundata*) and that could be described and predicted by mathematics.[13]

Leibniz and Predictability

Hardy Grant in "Leibniz and the Spell of the Continuous" (1994) argues that Leibniz's mathematics reflected his deep belief in predictability and his denial of chaos. "The order and predictability that [Leibniz] saw everywhere in the nature of things, the absence of chaos and caprice, were gifts of a benevolent God, and the source of the world's perfection and beauty." Moreover, Grant sees Leibniz as rejecting assumptions about discontinuity and unpredictability that would later become fundamental to chaos theory.

> On at least one occasion Leibniz flirted briefly with the kind of spectacu-lar natural discontinuity now studied by chaos theorists: he pictured, as an example of a tiny cause with immense effects, a small spark destroying an entire city by igniting a quantity of gunpowder. But he dismissed such appar-ent anomalies as . . . outside the general rule.[14]

Although Leibniz did not use the terms *natura naturans* or *natura naturata*, he was nevertheless deeply engaged with the issues underlying these concepts. His mature philosophy was founded on living forces (*vis viva*) and well-founded phenomena (*phenomena bene fundata*). His mathematical framework described the phenomenal world of *natura naturata*. The unfolding, active lives of his windowless monads characterized the world of *natura naturans*.

In "The Monadology" and "The Principles of Nature and Grace," written in 1714 two years before his death, Leibniz set out his system of physics and metaphysics. At the root of his philosophy were the laws of noncontradiction and sufficient reason. Substance comprised an infinity of active, point-sources of force, or monads, each with a striving or *conatus* toward a future state. Each monad was not only windowless, meaning that nothing external could enter into it, but it also mirrored the entire universe. The lives of all the monads unfolded together in a system of preestablished harmony created at the beginning by God. The world that God created was chosen by Him from an infinite number of possibilities so that the world that He actually chose to create was the best of all. The real world—the best of all possible worlds—was thus dynamic, active, and alive. Force was the foundation of the universe.[15]

The phenomenal world of everyday life comprised corporeal objects that were collections of monads (or minds) in confused states and perceived as extended bodies. Corporeal bodies had extension, size, shape, and motion and were "well-founded" in the simple monads constituting them. Whereas force was real, the very essence of the monad, corporeal bodies were perceived as having extension and motion as attributes rather than substance. Bodies in motion obeyed the law of conservation of living force, or *vis viva*, the product of its mass and the square of its velocity (the distance through which it fell under acceleration). The body's *vis viva* or mv^2 (later with the addition of ½ called kinetic energy) was conserved in elastic impacts, whereas in semi-elastic and inelastic impacts it was held in the body's small parts and not lost to the universe.[16]

Leibniz's contribution to mechanism was dynamics, the foundation of the general law of conservation of energy. By demonstrating the validity of the mass times the square of the velocity (mv^2), he influenced a generation of Enlightenment thinkers from Gabrielle Emile du Châtelet to Jean d'Alembert and beyond. By his discovery of the calculus, he contributed to the predictability and control of nature. Mathematics and mechanics were the basis of rationalism. Together they described the phenomenal world of *natura naturata*.

His contribution to metaphysics was the idea that the real world was active and alive. The creativity historically associated with *natura naturans* was manifested in God's infinite wisdom and foresight in producing, at the beginning of time, all the monads whose lives unfolded together in a pre-established harmony. Through his rationalism, Leibniz, like Spinoza, rejected the disorderly, chaotic aspect of *natura naturans*. Seemingly chaotic events were simply part of the pre-ordained unfolding coordination of all the events in the real and phenomenal worlds such that what actually came to pass was the best of all possible worlds.

Long after Spinoza's death, and after Leibniz had worked out his mature system in the *Theodicy* (1710) and "The Monadology" (1714), Leibniz wrote to a correspondent:

132   *Controlling Nature*

[It] is precisely by means of the monads that Spinozism is destroyed. For there are as many true substances—as many living mirrors of the Universe, always subsisting, as it were, or concentrated Universes—as there are Monads; whereas, according to Spinoza, there is but one sole substance. He would be right, if there were no Monads.[17]

In other words, Leibniz admits, "if I were not right, he would be." From a modern perspective, the rationalism of Spinoza, which declared a pantheistic world of nature as the ultimate reality, and the vitalism of Leibniz based on a living, active nature were both harbingers of an ecological worldview. In another way, Newton also made contributions to rationalism, predictability, and control, as well as to active principles in nature. And yet, just as Leibniz and Spinoza had harbored deep disagreements, so Leibniz and Newton were deeply at odds about the foundations of rationalism. They came to blows over the nature of God, mechanics, and mathematics.

Newton

Perhaps the foremost achievement of Isaac Newton's (1642–1727) *Philosophiae Naturalis Principia Mathematica* (*Mathematical Principles of Natural Philosophy*, 1687) was that by reducing the observable universe to the law of gravitation and three laws of motion, predictions could be made from current data. From an initial set of observations of celestial and terrestrial phenomena at a given moment, future states could be predicted. Newtonian science, in its public form, in the *Principia Mathematica* engages with the meaning of what had been called *Natura naturata*—the created world—in ways that promote predictability and reject chaos.[18]

Prophetically, Isaac Newton was born on Christmas Day, 1642, the same year that the English Civil War was breaking out and Galileo died. The times were both politically turbulent and intellectually challenging. The previous year Descartes had published his *Meditations on the First Philosophy* arguing not only that "I think therefore I am" (*cogito ergo sum*), but that God, as a clear and distinct idea, must therefore exist. The following year Louis XIV, the Sun King, ascended to the throne of France at the age of five, and Italian physicist Evangelista Torricelli (1608–1647) replaced his water barometer (consisting of a 35-foot tube protruding from the roof of his house) with a one-meter tube filled with mercury. Whether the space at the top of the barometer was a vacuum or a plenum held enormous implications for questions of void space, atomism, and God's immanent presence in the world versus his action at a distance.

Ironically, Newton's *anni mirabiles* of 1665–1666 that led to the 1687 *Principia Mathematica* occurred during a period of chaos in London, brought about by the advent of the Bubonic Plague of 1665 and the great fire of London in 1666. After receiving his degree at Trinity College, Cambridge, in August 1665, with the university now closed, Newton retreated to his home in Woolsthorpe outside London. It was there that he laid the foundations for his theories of the calculus, optics, and law of gravitation, returning to Cambridge in 1667 as a fellow of Trinity College. In 1669 he was appointed Lucasian professor at Cambridge, but it was not until 1687 that his major synthesis of the laws of motion and gravitation was published as the *Principia Mathematica*, a work that advanced the possibility of predicting and controlling nature.

Laws of Nature 133

Figure 6.4 The Great Plague of London, 1665.
The last major outbreak of the Bubonic Plague in England.
Artist unknown.

Barbara Adam in *Timescapes of Modernity* (1998) characterizes Newton's mechanistic science and its differences from chaos theory in the following way:

> In Newtonian science there is no room for uniqueness and creativity, contingency and contextuality, surprises and discontinuities, chaos and catastrophes. . . . In many ways, chaos theory inverts Newtonian logic. Instead of abstraction and linear cause-and-effect chains, [chaos] theory focuses on

134 *Controlling Nature*

interdependent, contingent connectivity. Instead of certainty and predictability it provides a rationale for the impossibility of establishing connections backwards from symptoms to cause or forward from action to future outcome. Where Newtonian science assumes single parts in motion, chaos theory presupposes complexity and associates this with continuous iterative interaction (that is, where the repetition of actions involves feedback).[19]

For Newton, the mechanical philosophy comprised four fundamental concepts—matter, motion, force, and void space. These built on and modified the system of matter and motion within a plenum that had been advocated by Descartes and Hobbes. Impressed force acted directly on a body to change its state of rest or motion in a straight line and was expressed as the force equals a body's mass times its acceleration (in modern terms, $F = ma$). Gravitational force acted at a distance, attracting all particles of matter toward each other according the inverse square of the distance between them (in modern terms, $F = G\, m_1 m_2 / r^2$, where G is the gravitational constant, the value of which was determined more than a century later). In the case of planetary motion, the force acts as if it were concentrated at the center of the planet or the earth. To Newton's disappointment, however, he was unable to provide a mechanical explanation for the way in which gravitational forces acted. (All proofs used Euclidean Geometry.) In his General Scholium to the second edition of the *Principia* (1713), he wrote:

> I have not yet been able to discover the cause of those properties of gravity from phenomena, and I frame no hypotheses. . . . [It] is enough that gravity does really exist, and act[s] according to the laws we have explained, and abundantly serves to account for all the motions of celestial bodies. . . . [20]

The mechanical philosophy created a dualism between passive matter and external force. From Aristotle's framework of four causes of motion—material, formal, efficient, and final, it eliminated the qualitative aspect of the formal cause and the final cause (*telos*). What remained were the material cause (matter), the quantitative aspect of the formal cause, and the efficient cause (force). Of the formal cause, only quantities, i.e., primary qualities, were real, objective. Secondary qualities—perceptions of sight, sound, touch, taste, and smell, generated by matter in motion (quantities)—were the subjective effects of the primary qualities on our human organs of sense. Without the mathematical dimension of the formal cause, the world would be chaotic, resembling the chaos (*anake/chora—Necessity/ Receptacle*) that existed before Plato's *Demiurge* organized it in accordance with the pure forms (see Chapter 1). Mathematical equations held in an ideal world of point sources of force, frictionless planes, perfect spheres, and a vacuum (no air resistance). As a quantitative science, mechanism eliminated teleological and qualitative explanations relating to hierarchies, values, purposes, harmonies, and qualitative properties. It emphasized quantities, structure, stability, and identity rather than flux, becoming, and process.[21]

Laws of Nature 135

Matter, for Newton, was divided into "hard massy particles" separated by void space. These particles comprised both matter and light and varied in size, shape, and weight. In the *Opticks*, published in four editions between 1704 and 1730, Newton wrote:

> God in the beginning formed matter in solid, massy, hard, impenetrable moveable particles, of such sizes and figures and with such other properties and in such proportion to space as most conduced to the end for which he form'd them. . . . [22]

Although Newton's *Principia* became the model for the mechanical worldview and an inspiration for future inquiries into astronomical and terrestrial phenomena, Newton himself was deeply concerned about the fact that the universe was apparently running down. It seemed to him that passing comets disrupted the orbits of planets, requiring periodic repair by God and the recruitment of new motion to offset evident decay. In Query 31 of the *Opticks*, he stated that the world could not have arisen from chaos merely by mechanistic principles of nature (as Descartes had implied in his 1644 *Principles of Philosophy*), but was created by an intelligent deity. Nevertheless irregularities could occur owing to mutual gravitational interactions among planets and comets requiring God's periodic restoration:

> Now by the help of these Principles, all material Things seem to have been composed of the hard and solid Particles above-mention'd, variously associated in the first Creation by the Counsel of an intelligent Agent. For it became him who created them to set them in order. And if he did so, it's unphilosophical to seek for any other Origin of the World, or to pretend that it might arise out of a Chaos by the mere Laws of Nature; though being once form'd, it may continue by those Laws for many Ages. For while Comets move in very excentrick Orbs in all manner of Positions, blind Fate could never make all the Planets move one and the same way in Orbs concentrick, some inconsiderable Irregularities excepted, which may have risen from the mutual Actions of Comets and Planets upon one another, and which will be apt to increase, till this System wants a Reformation. [23]

In addition to occasional repairs to the cosmic system, it seemed, moreover, that certain kinds of processes could not be explained by the laws of motion and gravitation as set out in the *Principia*. Some other rational principles lay behind the phenomena he had successfully submitted to mathematical analysis.

In the Queries to the *Opticks*, Newton concerned himself with the precise question of violent, seemingly chaotic motions in the cosmos. In Query 31, he characterized the earth as permeated with hot springs, burning mountains, earthquakes, hurricanes, exploding fires, tempests, landslides, boiling seas, and thunderous rains. Acid vapors caused fermentations that replenished the earth with

136   *Controlling Nature*

new motions in violent clashes. Fermentations could cause particles at rest to be "put into new motions by a very potent Principle, which acts on them only when they approach one another and causes them to meet and clash with great violence. . . . "[24]

Such chaotic motions reflected the unruly dimension of creative nature (*natura creans/naturans*) about which earlier philosophers had speculated. But far from implying the recalcitrance of a personified nature, the acts of a vengeful God, the soul of the world, or even the active body of the world, activity for Newton stemmed from God's active principles. God both created the cosmos and by his will continually sustained it. In Query 31 of the *Opticks*, he wrote,

> And yet we are not to consider the world as the body of God, or the several parts thereof, as the Parts of God. He is a uniform Being, void of Organs, Members or Parts, and they are his Creatures subordinate to him, and subservient to his Will; and he is no more the Soul of them, than the Soul of Man is the Soul of the Species of Things. . . . And God has no need of such Organs, he being every where present to the Things themselves.[25]

What was required for God's continual action and presence in the world, Newton believed, was a naturalistic explanation—stemming from what he called "active principles." These "active principles," such as the cause of gravity and the cause of fermentation, were the means by which new motion was recruited to keep the universe from running down. Had Newton used the term *Natura naturans*, he might have applied it to God's continual presence as manifested through active principles. He discussed these ideas in the Queries to the *Opticks*, in the form of speculative observations on the nature of God.

In the Query 31, he conjectured that violent clashes within the earth were created through active principles, such as the causes of fermentation, gravity, and cohesion—principles that were naturalistic rather than miraculous. He speculated that the attractions of gravity, magnetism, and electricity were the result of impulses or other means that drew them toward each other or caused them to fly apart. He asked, "Have not the small Particles of Bodies certain Powers, Virtues, or Forces, by which they act at a distance. . . . "? Bodies were not only attracted toward each other, but also had repulsive forces "by which they fly from one another."[26]

Concerned that the total motion was decreasing and the universe would ultimately decay and run down, he argued that

> [T]here is a necessity of conserving and recruiting it by active Principles, such as the cause of Gravity, by which Planets and Comets keep their Motions in their Orbs, and Bodies acquire great Motion in fall; and the cause of Fermentation by which the Heart and Blood of Animals are kept in perpetual Motion and Heat; the inward Parts of the Earth are constantly warm'd, and in some places grow very hot; Bodies burn and shine, Mountains take fire, the Caverns of the Earth are blown up, and the Sun continues violently hot and lucid, and warms all things by his Light.[27]

Thus for Newton new motion had to be continually created to offset decay, whereas for Leibniz activity (force) was put into the world at the creation and was continually maintained within the unfolding lives of the monads. For Newton, these active principles were an expression of God's continual care and intervention in the world. He wrote:

> For we meet with very little Motion in the World, besides what is owing to these active Principles. And if it were not for these principles, the Bodies of the Earth, Planets, Comets, Sun, and all things in them, would grow cold and freeze, and become inactive Masses; and all Putrefaction, Generation, Vegetation and Life would cease, and the planets and Comets would not remain in their Orbs.[28]

For Leibniz, however, such principles were a limitation on God's logic and sufficient reason. The differences between the two scientists came out forcefully in the debates of 1716 that occurred just before Leibniz's death.

Leibniz and Newton on the Nature of God

The debate between Newton and Leibniz over the nature of God was carried out in 1715–1716 in the Leibniz-Clarke correspondence, with theologian Samuel Clarke (1675–1729) speaking for Newton. The debate centered on the issue of the primacy of God's will versus his logic—was God primarily a voluntarist or an intellectual deity? For Newton, the existence of the world depended on the will of God. His supreme power was demonstrated by his continual presence and the interposition of his will. This theological position was consistent with voluntarism, or the idea that God's will revealed his glory and providential care. For Newton and Clarke, God created the world and through his will continued to sustain it.[29]

In the General Scholium to the *Principia mathematica* (1713), Newton wrote: "The Supreme God, is a Being eternal, infinite, absolutely perfect, but a being, however perfect, without dominion, cannot be said to be 'Lord God'. . . . " Here, according to Edward Davis, "divine perfection was virtually equated with dominion, which he understood to be manifest in the constant activity of the divine will."[30]

For Leibniz, on the other hand, God's logic and reason superseded his will. God created the world at the beginning of time as a logically perfect world that required no further tinkering or intervention. The need to intervene would mean that God was an imperfect deity. For Leibniz, the world was the best possible world because it resulted from his logic; moreover, because of the principle of sufficient reason, it was the best of all the possible worlds he could have created. Owing to our limited understanding we are not privy to the reasons for its apparent imperfections and evils.

Closely linked to the character of God was the character of the machine-like model of the world. The world of many seventeenth-century natural philosophers was like a clock, or a well-oiled machine made up of inert parts—dead corpuscles or atoms that had no spirit or life within them. Did God simply wind up his clock

138   Controlling Nature

and let it tick away into eternity, as the Newtonians accused the Leibnizians of arguing. Or did God continually repair his machine or oil his clock to prevent it from breaking down? Was that clock like a real-world pendulum clock that slowed down owing to the friction of its point of contact or was it an ideal pendulum ticking away forever in an ideal, frictionless world? One's answer to these fundamental questions depended on one's vision of God and nature.

Leibniz began his debate with Samuel Clarke in a letter of November 1715 in which he wrote:

> Sir Isaac Newton and his followers have also a very odd opinion concerning the work of God. According to their doctrine, God Almighty wants [i.e. needs] to wind up his watch from time to time; otherwise it would cease to move. He had not, it seems, sufficient foresight to make it a perpetual motion. Nay, the machine of God's making is so imperfect according to these gentlemen that he is obliged to clean it now and then by an extraordinary concourse, and even to mend it as a clockmaker mends his work, who must consequently be so much the more unskillful a workman, as he is oftener obliged to mend his work and to set it right.[31]

Clarke's response continued the clock metaphor, arguing that God was not like an ordinary clockmaker, who creates only the gears and wheels, but cannot create the force itself. God Almighty, however, is both the creator and sustainer of all motions and maintains them throughout time. "The notion of the world's being a great machine," he stated,

> going on without the interposition of God, as a clock continues to go without the assistance of a clockmaker, is the notion of materialism and fate and tends . . . to exclude providence and God's government in reality out of the world.[32]

Leibniz answered by setting out the differences in God's power and wisdom. Yes, God's power was demonstrated by the fact that he created the world. Only a deity with infinite power could do so. But to create this world and no other, another principle was required, that of wisdom. Once again asserting his differences with both Spinoza and Newton, Leibniz wrote:

> Thus the skill of God must not be inferior to that of a workman; nay, it must go infinitely beyond it. The bare production of everything would indeed show the power of God, but it would not sufficiently show his wisdom. They who maintain the contrary will fall exactly into the error of the materialist and of Spinoza, from whom they profess to differ. They would, in such case, acknowledge power, but not sufficient wisdom, in the principle or cause of all things. [33]

God thus had perfect foresight, having anticipated and provided the best possible outcome for all contingencies when he created this world and no other.

Laws of Nature 139

Were Newton's active principles by which new motion was put into the universe actually miracles? According to Leibniz, they were indeed, especially the cause of gravity. Reiterating a critique he had first made in 1712, Leibniz wrote to Clarke that

> If God could cause a body to move free in the ether round about a certain fixed center, without any other creature acting upon it, I say it could not be done without a miracle, since it cannot be explained by the nature of bodies.

In other words, since for Newton, the cause of gravity was not known, it must be called a miracle.[34]

In a draft response to Leibniz's 1712 criticism, Newton had argued that gravity should not be called a miracle just because no cause had been found for it:

> If any man should say that bodies attract one another by a power whose cause is unknown to us or by a power seated in the frame of nature by the will of God . . . I know not why he should be said to introduce miracles. . . . "

That is, we may not yet know the cause of gravity, but that does not mean it is a miracle. He argued further that gravity does not require a miracle. He wrote: "[C]ertainly God could create Planets that should move round of themselves without any other cause [than] gravity. . . . For gravity without a Miracle may keep the Planets in." In other words, God can create planetary motion without recourse to a miracle. All causes or actions may not have mechanical explanations, but that does not mean they are miracles.[35]

The cause of gravity was one of Newton's active principles that were needed to keep the universe from running down. The others were the causes of fermentation and cohesion. But whether these were material or immaterial, he cannily left to his readers to decide. In a letter to British theologian Richard Bentley in 1693, Newton wrote: "Gravity must be caused by an agent acting constantly according to certain laws, but whether this agent be material or immaterial is a question I have left to [the] consideration of my readers."[36]

Newton's unwillingness to state whether the cause of gravity and other active principles were material or immaterial left open to debate the character of these immaterial causes. Did they stem from God's will, hence he could use his providential care to keep the universe from running down? Or, more dramatically, did God create miracles or willfully or even vengefully intervene in the world? Or, third, were they natural causes that had not yet been explained by natural philosophy and mathematics, hence did not require a deity? The first question led to further investigations on whether the universe was running down and ultimately to the development of classical thermodynamics. The second to whether chaotic, seemingly unpredictable events, such as earthquakes and volcanoes, were examples of God's vengeance or could they be explained by natural philosophy. The third raised the question of whether God was a necessary hypothesis nor whether he was needed at all (as some would later argue).

140   *Controlling Nature*

Implications of the Leibniz-Newton Debate

Through their mutual concerns over activity and decay in the universe, both New-ton and Leibniz, in a sense, anticipated later developments in classical thermody-namics. Leibniz's law of the conservation of *vis viva* evolved into the general law of the conservation of energy (the first law of thermodynamics) which holds that the total energy in the universe can neither be created nor destroyed; it is only changed in form (as it is transferred from radiant to chemical to hydrodynamic to meta-bolic energy, and so on). And Newton, in another sense, anticipated the second law of thermodynamics, which holds that the energy available for work (the useful energy) in a closed system (such as a steam engine) is decreasing. The total entropy of the system (the energy unavailable for work) is ever increasing. The universe as a whole is running down (as Newton had feared), just as a clock runs down when it is not rewound. The second law implies that the universe is moving from order to disorder, hot objects become cold, living things die, people grow older, and that in billions of years the universe as a whole will reach a uniform temperature.

The second law of thermodynamics, as formulated in the nineteenth century, applied to closed, stable systems that are in equilibrium or near-equilibrium, such as pendulum clocks, steam engines, refrigerators, and solar systems. Small inputs lead to adjustments and adaptations. But in the late twentieth century, Ilya Prigogine (1917–2003) argued that, in fact, new order can emerge from disor-der in far-from-equilibrium thermodynamics. In far-from-equilibrium systems, inputs are so large that the system cannot adjust and new nonlinear relationships take over. New levels of organization can emerge out of disorder when a system breaks down. Prigogine's approach applies to open rather than closed systems, such as ecological and social organizations. In open systems, matter and energy are exchanged with the environment. In biological systems, new cellular structures or life forms can emerge; in social systems, societies can reorganize around new economic forms such as the change from hunting-gathering to horticulture or from preindustrial to industrial economies.

The emergence of chaos and complexity theories in the late twentieth century means that the laws of mechanics discovered by Newton and Leibniz apply only to a limited domain that can be described by solvable, deterministic, predictive math-ematics, using calculus and first-order linear differential equations, along with sta-tistics and probability theory. For chaos theory, by contrast, the usual domain is that of open, nonlinear systems described by probabilities and higher-order differ-ential equations in which solutions can only be approximated and where predic-tive capacities are limited. Such situations include unusual weather and geological events whose mechanics may be understood, but which may nevertheless occur suddenly and without warning.

The Lisbon Earthquake

On November 1, 1755, "All Saint's Day," at 9:30 A.M., as if in defiance of Leibniz and Newton's rational system of mathematical predictability, an earthquake with a probable strength of 9.0 on the (subsequent) Richter scale, unexpectedly struck

Lisbon, Portugal. It lasted a full ten minutes and was felt as far away as North Africa, southwest Spain, France, Switzerland, and northern Italy. It destroyed much of Lisbon, spawned a tsunami, and caused several severe fires that lasted for five days. Churches, mosques, synagogues, castles, hospitals, palaces, opera houses, aqueducts, and houses collapsed. Debris-filled streets and raging fires impeded escape. Those who lived along the coast or tried to leave by boat were swept away by the tsunami. Boats capsized and sunk, buildings were flooded, and city walls destroyed. Within a few hours, the tsunami had traveled across the Atlantic to the West Indies.[37]

The causes of earthquakes had excited considerable speculation over the centuries. Explanations included those of Epicurus (341–270 B.C.E.) who wrote that "Earthquakes may be brought about because wind is caught up in the earth, so the earth is dislocated in small masses and is continually shaken, and that causes it to sway" and Lucretius (1st c. B.C.E), who noted that during a quake, "The earth will lean and then sway back, [i]ts wavering mass restored to the right poise. That explains why all houses reel, top floor [m]ost, then the middle, and ground floor hardly at all."[38]

As discussed above, seventeenth-century natural philosophers sought rational explanations for earthquakes and other seemingly chaotic events. Spinoza held that nothing happens in nature which does not follow from nature's laws and denied that such events were miracles, insisting that they were irrationally ascribed to God's action. For Leibniz, to whom God's intellect was prior to his will, unpredictable events were part of God's preestablished harmony and the best of all possible worlds. Apparent evils in our limited world of experience were imperfections only if the larger picture to which we are not privy is not viewed. For Newton, although God's will was prior to his intellect, his actions were not willful acts of retribution. Other causes were needed to explain unpredictable events. Those stemmed from active principles by which new motion was recruited in a world in danger of running down—such as "the cause of gravity, by which planets and comets acquire great motion in fall; and the cause of fermentation, by which the heart and blood of animals are kept in perpetual motion and heat. . . . "[39]

Other seventeenth-century scientists who sought naturalistic explanations included Robert Hooke (1635–1703). In his *Lectures and Discourses of Earthquakes* (1668, published posthumously in 1705), Hooke wrote that during the "Eruption of some kind of Subterraneous Fires or Earthquakes, great quantities of Earth have been deserted by the Water and laid bare and dry." Hooke's law, "A True Theory of Elasticity and Springiness" (1676), which stated that the force needed to compress or extend a spring by a particular distance is proportional to that distance ($f = -kx$, where k is the elastic constant of the spring) would later become the basis for understanding the action of earthquakes.[40]

Despite such efforts at rational explanations, however, theologians continued to ascribe them to the willful vengeance of God acting through nature in retribution for human sins. Jesuit priest Gabriel Malagrida (1689–1761), who was living in Lisbon at the time of the quake, immediately attributed it to the wrath of God. "Learn, O Lisbon," he warned,

Figure 6.5 The Lisbon Earthquake, 1755.
From G[eorg] Hartwig, *Volcanoes and Earthquakes: A Popular Description of the Movements in the Earth's Crust: From "The Subterranean World"* (London: Longmans Green, 1887), frontispiece.

> that the destroyers of our houses, palaces, churches, and convents, the cause of the death of so many people and of the flames that devoured such vast treasures, are your abominable sins, and not comets, stars, vapours and exhalations, and similar natural phenomena. . . . As for the dead, what a great harvest of sinful souls such disasters send to Hell! It is scandalous to pretend the earthquake was just a natural event. . . .

He further declared that a city imbued with such sinful ways should devote all its strength and energy to repentance. Malagrida's view was consistent with the teachings of the Middle Ages and Renaissance that God acting through nature as his vice-regent would exact payment for human sins in the form of famines, droughts, earthquakes, and volcanic eruptions.[41]

The Lisbon Earthquake also evoked a debate between Voltaire (1694–1778) and Jean Jacques Rousseau (1712–1778) on the cause of the earthquake and by implication the meaning of natural disasters. One of Voltaire's more notable diatribes against Leibniz's "best of all possible worlds" was a "Poem on the Lisbon Disaster, or: An Examination of that Axiom, 'All Is Well'" (1755). In it, Voltaire lampooned Leibniz's idea that such disasters were the preestablished result of eternal laws directing the best possible world. To blame the disaster on the impersonal laws of nature was absurd.

> Oh, miserable mortals! Oh wretched earth!
> Oh, dreadful assembly of all mankind!
> Eternal sermon of useless sufferings!

Deluded philosophers who cry, "All is well,"
Hasten, contemplate these frightful ruins,
This wreck, these shreds, these wretched ashes of the dead;
These women and children heaped on one another,
These scattered members under broken marble;
One-hundred thousand unfortunates devoured by the earth
Who, bleeding, lacerated, and still alive,
Buried under their roofs without aid in their anguish,
End their sad days!
In answer to the half-formed cries of their dying voices,
At the frightful sight of their smoking ashes,
Will you say: "This is result of eternal laws
Directing the acts of a free and good God!". . . . [42]

Unable to contain himself over what he saw as Voltaire's crass spoofing of Leibniz at the expense of the misery and suffering of real human beings, Jean-Jacques Rousseau, in 1756, penned a caustic response. Underlying the debate was the very meaning of a "natural disaster." Rousseau argued that it was not nature that created the disastrous outcome, but that the misery was humanly caused, resulting from the fact that it occurred in a humanly constructed city, containing hundreds of humanly constructed buildings where thousands of people lived. Had it occurred in a remote wilderness, it would not even have been noted. What was at stake was the misery of the human condition, not the laws of nature.

> All my complaints are . . . against your poem on the Lisbon disaster, because I expected from it evidence more worthy of the humanity which apparently inspired you to write it. You reproach Alexander Pope and Leibnitz with belittling our misfortunes by affirming that all is well, but you so burden the list of our miseries that you further disparage our condition. . . . [43]

The contrast between the religious interpretation of earthquakes as God's payment for sin and the Enlightenment interpretation that they arose from natural causes found expression in David Hume's pronouncement in 1748 that "a miracle is a violation of the laws of nature," and Baron Paul-Henri Thiry d'Holbach's 1770 proclamation that "experience teaches that Nature acts by simple, uniform, and invariable laws." The source of human misery, he maintained, is "ignorance of Nature" and adherence to the "blind opinions" of the past. Man "by his senses must penetrate her secrets" . . . and "draw experience of her laws."[44]

Yet, despite the movement away from religious retribution and toward naturalistic explanations, the Lisbon Earthquake was one of the more poignant examples of the limitations of the mechanistic paradigm. Notwithstanding remarkable efforts and considerable success in extending mechanistic explanations of nature to other fields, mechanism still could not predict with accuracy the occurrence of seemingly chaotic events such as geological catastrophes and severe weather disruptions. The unruly, defiant, and unlawful dimension of *natura naturans* remained to resurface in the chaos paradigm of the late twentieth century.

144 *Controlling Nature*

Conclusion

By the late nineteenth century, the mechanistic view of nature led to numerous scientific and technological advances. Enhanced by differential equations and probability theory, the new sciences of hydrodynamics, thermodynamics, and electromagnetism gave rise to optimism over the control of nature. Technologies such as dams and turbines allowed water to be captured and regulated for electricity and irrigation. Steam engines propelled steamboats and trains and created industrialized factories. Telegraphs and telephones enabled instant communication. By the end of the nineteenth century, it seemed to some that mechanistic physics had reached a state of near completion.

Newtonian mechanics, however, was challenged in the early twentieth century by Einstein's Theory of Relativity (1905) and by Heisenberg's Principle of Uncertainty (1927). Relativity theory introduced the relativity of simultaneous events at the level of very large velocities and quantum mechanics introduced uncertainties at subatomic dimensions. At very high velocities approaching the speed of light (relative to a "stationary" observer), clocks would slow down and meter sticks would contract. Therefore, events which appear simultaneous to one observer would not be simultaneous to another observer when the two are in relative motion with respect to each other. On the other hand, at subatomic (quantum level) dimensions, both the momentum and position of a particle cannot be known simultaneously.

In the 1970s, chaos and complexity theories went further, challenging predictability at the level of everyday life. The great triumph of mechanistic science had been its ability to predict. If one can solve equations, one can predict and therefore control and manage nature. But chaos and complexity theories argue that predictability pertains to a limited domain of nature. The larger domain is one of unpredictability.

It may be known that sometime in the next hundred years an earthquake of large magnitude will occur on a particular fault, but not when. Or that a small atmospheric perturbation in one place can result in a large atmospheric disturbance in another place, but not where. Or that human activities can impact climate, but with consequences that cannot be controlled. What then are the implications of "nature naturing" for the twenty-first century?[45]

Notes

1 Matthew Stewart, *The Courtier and the Heretic: Leibniz, Spinoza and the Fate of God in the Modern World* (New York: Norton, 2006), p. 15.
2 Stewart, *The Courtier and the Heretic*, p. 195.
3 Stewart, *The Courtier and the Heretic*, pp. 196–199, 217 (quotation).
4 On the successes of and models for idealized conditions that draw on Platonic idealism and Aristotelian essentialism, see Daniel Simberloff, "A Succession of Paradigms in Ecology: Essentialism to Materialism and Probabilism," *Synthese* 43, no. 1 (January 1980): 3–39: "Idealism views the material objects of the world as imperfect embodiments of fundamental, unchanging essences or ideal formal structures" (p. 3). "[T]here is something profoundly disturbing about a nature in which random elements play a

Laws of Nature 145

large role. . . . Greek metaphysics will not vanish easily" (p. 30). On the Lisbon Earthquake, see discussion below. Also http://en.wikipedia.org/wiki/1755_Lisbon_earthquake and http://geophysics-old.tau.ac.il/personal/shmulik/LisbonEq-letters.htm.

5 Stewart, *The Courtier and the Heretic*, pp. 12, 88–90, 140, 151–152.
6 Stewart, *The Courtier and the Heretic*, pp. 151–155.
7 Stewart, *The Courtier and the Heretic*, pp. 219–220.
8 Stewart, *The Courtier and the Heretic*, pp. 235–241.
9 Gottfried Wilhelm Leibniz, "Note, February 11, 1676," in *Philosophical Papers and Letters*, ed. Leroy Loemker, 2 vols. (Chicago: University of Chicago Press, 1956), vol. 1, p. 246.
10 Leibniz, "Dissertation on the Art of Combinations" (1666), in *Philosophical Papers and Letters*, ed. Loemker, vol. 1, pp. 117–133; Stewart, *The Courtier and the Heretic*, pp. 79–80.
11 Leibniz, "Letter to Henry Oldenburg," December 28, 1675, in *Philosophical Papers and Letters*, ed. Loemker, vol. 1, pp. 256–258, quotation on p. 258.
12 Leibniz, "Note, February 11, 1676," in *Philosophical Papers and Letters*, ed. Loemker, vol. 1, p. 246.
13 Leibniz, "The Monadology" (1714), in *Philosophical Papers and Letters*, ed. Loemker, vol. 2, pp. 1044–1061; Stewart, *The Courtier and the Heretic*, pp. 150–151.
14 Hardy Grant, "Leibniz and the Spell of the Continuous," *College Mathematics Journal* 25, no. 4 (September 1994): 291–294; Leibniz, *Philosophical Papers and Letters*, ed. Loemker, vol. 1, pp. 538–543, see pp. 539, 541.

> It is true that in composite things a small change can sometimes bring about a great effect. So a small spark, for example, which falls into a large mass of gunpowder can demolish an entire city. But this is not contrary to our principle, for it can be explained by these same general principles. Nothing like this can happen in primary or simple things, however, for otherwise nature would not be the effect of infinite wisdom. (Loemker, vol. 1, p. 541)

15 Leibniz, "The Monadology" (1714), in *Philosophical Papers and Letters*, ed. Loemker, vol. 2, pp. 1044–1061, sec. 31, 32, 7, 83, 53–54; idem, "The Principles of Nature and Grace" (1714), in *Philosophical Papers and Letters*, ed. Loemker, vol. 2, pp. 1033–1043, secs. 1–4, 12, 15; Leibniz, *Theodicy*, trans. E.M. Huggard (Gutenberg eBook, 2005), pp. 67, 128.
16 Leibniz, "Specimen Dynamicum" (1695), in *Philosophical Papers and Letters*, ed. Loemker, vol. 2, pp. 711–738, see pp. 712, 717.
17 Stewart, *The Courtier and the Heretic*, quotation on p. 278.
18 Isaac Newton, *Mathematical Principles of Natural Philosophy* (*Philosophiae Naturalis Principia Mathematica*), trans. Andrew Motte (1729), rev. Florian Cajori (Berkeley: University of California Press, 1960); The *Principia* was published in three editions: 1687, 1713, and 1726. Isaac Newton, *Opticks*, based on 4th ed. 1730 (New York: Dover, 1952). The Latin edition of the *Opticks* in 1706 contained twenty-three queries. To the 1717 edition, eight new queries were inserted and numbered 17–24. Query 23 of the 1706 edition appears as Query 31 in the 1717 edition. Citations refer to the 4th edition (1730); (repr., New York: Dover, 1952).
19 Barbara Adam, *Timescapes of Modernity: The Environment and Invisible Hazards* (New York: Routledge, 1998), Ch. 1, "Nature Re/constituted and Re/conceptualized," pp. 42–43, 46, 50–53, quotations on p. 43, 46.
20 Newton, *Mathematical Principles of Natural Philosophy*, 1713 edition, op cit., General Scholium, pp. 543–547, quotation on p. 547; Richard S. Westfall, *The Construction of Modern Science: Mechanisms and Mechanics* (New York: Wiley, 1971); Carolyn Merchant, *The Death of Nature: Women, Ecology, and the Scientific Revolution* (San Francisco: HarperCollins, 1980; 2nd ed. 1990), pp. 283–287.

146   *Controlling Nature*

21   Simberloff, "Succession of Paradigms in Ecology," op. cit.,

> The nineteenth century was dominated by a deterministic mechanics, hypostatized by twin hypothetical ideal beings, Laplace's Demon and Maxwell's Demon. The former could predict in Newtonian, cause-and-effect, action-reaction fashion the complete state of the universe, given knowledge of the positions and velocities of all its particles for a single instant. The latter could violate the second law of thermodynamics, and in so doing, construct a perpetual motion machine. (p. 10)

22   Isaac Newton, *Opticks*, 4th ed. (1730), Query 31, p. 400.

23   Newton, *Opticks*, Query 31, p. 402.

24   Newton, *Opticks*, Query 31, p. 380.

25   Newton, *Opticks*, Query 31, p. 403.

26   Newton, *Opticks*, Query 31, pp. 375–376, quotations on pp. 375, 387.

27   Newton, *Opticks*, Query 31, p. 399.

28   Newton, *Opticks*, Query 31, pp. 399–400.

29   Peter Harrison, "Was Newton a Voluntarist?" in *Newton and Newtonianism: New Studies*, ed. James E. Force and Sarah Hudson (Dordrecht: Kluwer, 2004), pp. 39–63, see esp. pp. 60–61; R[oger] S. Woolhouse, *Starting with Leibniz* (London: Continuum, 2010), "Evil in the Best Possible World," pp. 143–148. In a manuscript of around 1672, entitled, "Of Natures Obvious Laws and Processes in Vegetation," Newton wrote: "The world might have been otherwise than it is (because there may be worlds otherwise framed than this). . . . [T] the whole series of causes might from eter[n]ity have been otherwise here, because they may be otherwise in other places," quoted in Betty Jo Teeter Dobbs, *The Janus Faces of Genius: The Role of Alchemy in Newton's Thought* (New York: Cambridge University Press, 1991), p. 266. See also Newton, *Opticks*, Query 31, p. 404: God is able to "vary the laws of nature and make worlds of several sorts in several parts of the universe."

30   Newton, *Principia mathematica*, op. cit., General Scholium, p. 544. Edward Davis, "Newton's Rejection of the 'Newtonian World View': The Role of Divine Will in Newton's Natural Philosophy," *Science and Christian Belief* 3, no. 1 (1991): 103–117, quotation on p. 106. See also Davis, p. 117: "Divine will had ordered the universe and would renew it from time to time as he saw fit. Natural laws were actively imposed by that will. . . ."

31   Leibniz, "The Controversy Between Leibniz and Clarke" (1715–1716), in *Philosophical Papers and Letters*, ed. Loemker, vol. 2, pp. 1095–1169, quotation on p. 1096.

32   Samuel Clarke, "Controversy Between Leibniz and Clarke," in *Philosophical Papers and Letters*, ed. Loemker, vol. 2, p. 1098.

33   Leibniz, "Controversy Between Leibniz and Clarke," in *Philosophical Papers and Letters*, ed. Loemker, vol. 2, pp. 1101–1103, quotation on p. 1102.

34   Leibniz, "Controversy Between Leibniz and Clarke," in *Philosophical Papers and Letters*, ed. Loemker, vol. 2, "Mr. Leibniz's Third Paper," see p. 1112.

35   Davis, "Newton's Rejection of the 'Newtonian World View,'" quotations on p. 115; Newton wrote on the meaning of a miracle in an unpublished paper (Davis, p. 114, and note 44):

> For Miracles are so called not because they are the works of God but because they happen seldom & for that reason create wonder. If they should happen constantly according to certain laws imprest upon the nature of things, they would no longer be wonders or miracles, but might be considered in Philosophy as part of the Phenomena of Nature . . . notwithstanding that the cause of their causes might be unknown to us.

Davis concludes that

> theology exerted a subtle but significant influence on 17th century science, driving thinkers such as Newton to reject what they perceived to be the presumptuous claims of continental rationalism, the very sorts of claims that would be wrongly associated with his name. (p. 117)

Rosalind W. Picard, however, in "Newton: Rationalizing Christianity or Not," (http://web.media.mit.edu/~picard/personal/Newton.php), argues that Newton did in fact believe in miracles:

> The word "miracle" derives from the Latin verb "mirari" to create wonder or astonishment. My understanding (from Davis, 1991) is that Newton believed that God does ALL things in nature, whether usual or unusual. The ones done by God's established laws tend to be usual, and we consider these natural. The ones done seldom, without laws, tend to arouse wonder in us, and thus are termed "miracles."

36 Quoted in Davis, "Newton's Rejection of the 'Newtonian World View,'" p. 116; see also, "Original Letter from Isaac Newton to Richard Bentley," 189.R.4.47, ff. 7–8, Trinity College Library, Cambridge, UK, published online, October 2007: http://www.newtonproject.sussex.ac.uk/view/texts/normalized/THEM00258.

37 On the Lisbon Earthquake see T[homas] D[owning] Kendrick, *The Lisbon Earthquake* (London: Methuen, 1956); Merry Weisner, Julius R. Ruff, and William Bruce Wheeler, *Discovering the Western Past* (Boston: Houghton Mifflin, 1989), pp. 28–56; Theodore E. D. Braun and John B. Radner, *The Lisbon Earthquake: Representations and Reactions* (Oxford, UK: Voltaire Foundation, 2005).

38 Epicurus, "Letter to Pythocles," in *Epicurus: The Extant Remains*, trans. Cyril Bailey (Oxford: Clarendon Press, 1926), p. 71; Lucretius, *De Rerum Natura* (1st c. B.C.E.), trans. W. E. Leonard (New York: E. P. Dutton, 1950), Bk. 6, lines 558–577.

39 Baruch Spinoza, *Theological-Political Treatise*, trans. Michael Silverthorne, ed. Jonathan Israel (New York: Cambridge University Press, 2007), Ch. 6, "On Miracles," sec. 1, pp. 81–82; Leibniz, *Theodicy* as discussed in Woolhouse, *Starting with Leibniz*, see pp. 143–148 on the Lisbon Earthquake and evil in the best possible world; Newton, *Opticks*, 4th ed., 1730, Query 31, quotation on pp. 399–400; Merchant, *Death of Nature*, p. 287.

40 Robert Hooke, *Lectures and Discourses of Earthquakes* (1668), in *The Posthumous Works of Robert Hooke*, containing his Cutlerian Lectures and other Discourses read at the Meetings of the Illustrious Royal Society, published by Richard Waller (London: Printed by Sam Smith and Benj. Walford, [printers to the Royal Society] at the Princes Arms in St. Paul's Church-yard, 1705), pp. 320–321. On the relation of Hooke's law to the physics of earthquakes, see "History of Seismology," http://www.asc1996.com/history1.htm:

> The variation in energetic state of a biatomic molecule can be described in a simple model in which the bond joining the two atoms vibrates, so that the energy of the bond varies as the length of the bond varies. The change in energy with bond length is given by Hooke's Law: A True Theory of Elasticity or Springiness.

See also David Kubrin, "'Such an Impertinently Litigious Lady': Hooke's Great Pretending vs. Newton's *Principia* . . .," in Norman J. W. Thrower, ed., *Standing on the Shoulders of Giants: A Longer View of Newton and Halley* (Berkeley: University of California Press, 1990), pp. 55–90.

41 Gabriel Malagrida, "An Opinion on the True Cause of the Earthquake" (1756), in Kendrick, *Lisbon Earthquake*, p. 89.

42 Voltaire [François-Marie Arouet], *Oeuvres Completes de Voltaire*, nouvelle edition, vol. 9 (Paris: Garnier Freres, 1877), p. 470, trans. Julius R. Ruff, as quoted in Weisner et al., *Discovering the Western Past*, op. cit., pp. 44–45.

43 Excerpted from Theodore Besterman, ed., *Voltaire's Correspondence*, vol. 30 (Geneva: Institût et Musée Voltaire, 1958), pp. 102–115, trans. Julius R. Ruff, as quoted in Weisner et al., *Discovering the Western Past*, op. cit., p. 45. On Voltaire and Rousseau's differing cultural meanings of disaster, see Russell R. Dynes, "The Dialogue Between Voltaire and Rousseau on the Lisbon Earthquake: The Emergence of a Social Science View," *International Journal of Mass Emergencies and Disasters* 18, no. 1 (March 2000): 97–115, esp. pp. 106–108.

44 David Hume, *Essays: Moral, Political, and Literary* (Oxford, UK: Oxford University Press, 1963), pp. 519–521, 524–526, 540–541 as quoted in Weisner et al., *Discovering the Western Past*, op. cit., p. 48; Paul-Henry Thiry, Baron d'Holbach, *The System of Nature* (1770), trans. H.D. Robinson (Boston: J.P. Mendum, 1853), pp. viii–ix, 12–13, 15, 19–23, as quoted in Weisner et al., *Discovering the Western Past*, op. cit., pp. 50, 51.

45 Ilya Prigogine and Isabelle Stengers, *Order Out of Chaos: Man's New Dialogue with Nature* (New York: Bantam, 1984); James Gleick, *Chaos: Making a New Science* (New York: Viking, 1987); M. Mitchell Waldrop, *Complexity: The Emerging Science at the Edge of Order and Chaos* (New York: Simon and Schuster, 1992); Edward Lorenz, "Predictability: Does the Flap of a Butterfly's Wings in Brazil Set off a Tornado in Texas?" (1972), reprinted in Lorenz, *The Essence of Chaos* (Seattle: University of Washington Press, 1993); Jennifer Wells, *Complexity and Sustainability* (New York: Routledge, 2013).

Epilogue
Rambunctious Nature in the Twenty-First Century

> There is in all visible things an invisible fecundity,
> a dimmed light,
> a meek namelessness, a hidden wholeness.
> This mysterious Unity and Integrity is
> Wisdom, the Mother of all, *Natura naturans*.
>
> Thomas Merton[1]

The view from Yup'ik Eskimo Peter John's armchair, in Newtok, western Alaska, looks out on sheds drying salmon and herring. From it he can see the houses and buildings of his village's friends and relatives near the Ninglick River flowing toward the Bering Sea. Yet they somehow seem strange and out of place. They are tipping downward on their stilts along a wooden walkway that is itself on the verge of collapsing. Near the front of the village are wide open cracks in the land, several of which are some ten feet deep. A few feet below the surface are sparkling drips of melting permafrost. As Ed Pilkington described it on a visit to Newtok in 2008:

> It is this layer of melting ice that has turned Newtok into what one observer described as the Ground Zero of global warming. According to NASA, temperatures in Alaska have risen more than any other place on the planet in the past 50 years—by some 4F on average, and up to 10F in winter. The Arctic in general has experienced a rate of warming that is double the earth's average, in part as a result of what is known as positive feedback. The brilliant white surface of ice and snow normally reflects most radiation from the sun back into space. But once the ice starts to melt through warming temperatures, the exposed land absorbs the radiation, thus causing further warming and melting.[2]

Climate change is the twenty-first century's *marquee* exemplar of autonomous nature responding to humanly produced greenhouse gases. The burning of fossil fuels by humans triggers an increase in carbon dioxide, thus increasing its concentration as a dominant "greenhouse gas" in the upper atmosphere. That in turn radiates increased heat back to the earth causing the air, oceans, and land to warm. Nature as an autonomous actor responds through thermodynamic feedback loops, tipping points, and often unanticipated cascading effects. Here *natura naturans* (nature's creative forces) and *natura naturata* (the created world) interact

150   *Epilogue*

in complex, dynamic processes, many of which are beginning to have potentially irreversible effects on life on earth.

The problem of an unpredictable and potentially uncontrollable autonomous nature that engaged many of the best thinkers in the Western world has returned. The goal of managing and controlling nature through science and technology has even suggested to some that *the earth as we know it today* might be radically altered in the future (e.g., through climate change, genetically engineered species, nuclear apocalypse, and so on). The tensions between *natura creans/naturans* and *natura creata/naturata* have re-emerged in the light of chaos and complexity theories and new concepts of the interlinked relationships between human and nonhuman nature. The way in which nature as an autonomous system behaves depends on how humans behave in relationship to it.

In this chapter, I argue that in the twenty-first century we are living within a new paradigm based on autonomous nature. The mechanistic paradigm that reigned supreme during the eighteenth and nineteenth centuries (advanced by probability and statistics) was superseded by the revolution in relativity and quantum mechanics in the early twentieth century and in chaos and complexity theory in the late twentieth century. Mechanism and mechanistic explanations are still the norm, however, in molecular biology, genetics, neuroscience, parts of biochemistry, chemistry and physics, and parts of engineering and biophysics. While mechanistic explanations are still essential to many parts of science and pertain to much of the everyday world, humanity is experiencing a new mode of life on the planet. That new paradigm encompasses chaos and complexity theories as they emerged in the sciences in the 1970s and 1980s, along with new forms of politics, psychology, and ethics. The new chaos paradigm is based on *natura naturans* as the active, creating, but also unruly and unlawful nature that challenged lives and livelihoods from Greco-Roman times to the present. As Thomas Merton characterizes it in the epigraph to this chapter, *natura naturans* symbolizes "a hidden wholeness, a mysterious unity, a new wisdom." That new wisdom entails new ways to live in a warming world based on renewable energy, politics that regulate the use of fossil fuels and promote solar and wind, and an ethic of partnership with nature.

A New Paradigm for the Twenty-First Century

In 2012, historians of science commemorated the fiftieth anniversary of Thomas Kuhn's *Structure of Scientific Revolutions* (1962). In this landmark work, Kuhn introduced the term "paradigm" to describe "normal science" in particular fields of research, e.g., astronomy, physics, biology, and so on, that allowed scientists to continue to solve problems within the accepted laws and concepts in their fields. But when anomalies that did not fit within a field's normal science appeared, new paradigms emerged during periods known as scientific revolutions.[3]

Kuhn stated that

> scientific revolutions are inaugurated by a growing sense . . . often restricted to a narrow subdivision of the scientific community, that an existing paradigm

has ceased to function adequately in the exploration of an aspect of nature to which that paradigm itself had previously led the way.

A primary example was the shift from the Ptolemaic, earth-centered, closed cosmos of the medieval period to the Copernican, sun-centered universe of the early modern era. Culminating with the synthesis of astronomical and terrestrial mechanics in Newtonian gravitational theory and the laws of motion, a new view of nature as a machine that could be understood, managed, and repaired emerged.[4]

In the eighteenth and nineteenth centuries, mechanistic determinism was tempered by probability theory, statistics, and stochastic (or random) processes. Jacob Bernoulli (1655–1705) in his *Ars Conjectandi* (published posthumously in 1713) developed the law of large numbers, in which the greater the number of trials the more closely the average of random events gives a stable long-term result. Carl Friedrich Gauss (1777–1855), along with others, developed the theory of least squares and used it to predict the location in which the asteroid Ceres, discovered in 1801, would be observed when it emerged from behind the glare of the sun. Pierre-Simon Laplace (1749–1827), in his *Analytic Theory of Probabilities* (1812), extended the theory to fit data points to a line and to model error distributions. In the late nineteenth century, statistical mechanics developed by Ludwig Boltzmann (1844–1906), James Clerk Maxwell (1831–1879), and J. Willard Gibbs (1839–1903) helped to explain the random motions of particles in gases under variations in temperature. Probability and statistics were applied with great success not only in the natural sciences, but the social sciences, medicine, and public health.[5]

A second major paradigm shift was the revolution in relativity theory and quantum mechanics in the early twentieth century that applied to the highest velocity in the universe, that of light, and the very smallest subatomic dimensions with implications for the everyday world. These include nuclear reactors, lasers, transistors, and LED (light-emitting diode) lights.

Since the 1970s a third major paradigm shift has been underway in the sciences of chaos and complexity. These new sciences argue that mechanistic science holds only in the *unusual* domain wherein predictability is possible, but that the more *usual* domain includes weather and climate, along with open, interactive ecological systems that can be unpredictable. Chaos theory suggests that most environmental and biological systems, such as weather, noise, population, and ecological patterns, cannot be described accurately by the linear equations of mechanistic science and may instead be governed by nonlinear, chaotic relationships.[6]

James Gleick, in his Prologue to *Chaos: Making a New Science*, writes:

> The most passionate advocates of the new science [of chaos] go so far as to say that twentieth century science will be remembered for just three things: relativity, quantum mechanics, and chaos. Chaos, they contend, has become the century's third great revolution in the physical sciences. Just like the first two revolutions, chaos cuts away at the tenets of Newton's physics. As one physicist put it: "Relativity eliminated the Newtonian illusion of absolute space and time; quantum theory eliminated the Newtonian dream of a controllable

152   *Epilogue*

measurement process; and chaos eliminated the Laplacian fantasy of deterministic predictability."[7]

Such challenges to classical mechanistic physics were forcefully iterated by prize-winning physicists such as Edward Lorenz (1917–2008) of the Massachusetts Institute of Technology and Ilya Prigogine (1917–2003) of the University of Texas at Austin. To atmospheric physicist, Edward Lorenz, is attributed the now-famous "butterfly effect," or the concept of sensitive dependence on initial conditions. In his book, *The Essence of Chaos* (1993), he states that "an immediate consequence of sensitive dependence in any system is the impossibility of making perfect predictions." The atmosphere is "an intricate dynamical system" whose "irregularities are manifestations of chaos." Weather forecasting involves many irregularities and unanticipated outcomes that cannot be predicted in advance. He acknowledges that "For some real-world systems we even lack the knowledge needed to formulate the differential equations . . . that realistically describe surging waves, with all their bubbles and spray, being driven by a gusty wind against a rocky shore."[8]

Ilya Prigogine pushed the problem of predictability still further in his work on far-from-equilibrium thermodynamics, hydrodynamics, and many chemical processes in which irreversibility and nonlinearity can lead to bifurcations, meaning that the system could go in a number of new directions because nonlinear equations have several different solutions. The outcome therefore cannot be predicted with certainty. Unstable dynamic systems, such as the weather systems described by Edward Lorenz, behave differently than the stable planetary systems described by Isaac Newton. For Prigogine, instability holds implications for the evolution of new systems and fields.[9]

He writes:

> The main point I want to emphasize is that we see in nature the appearance of spontaneous processes which we cannot control . . . This is not giving up scientific rationality. After all, we have not chosen the world in which we are living. We scientists have to describe the world. . . . And the world in which we are living is highly unstable. What I want to emphasize, however is that this knowledge of instability may lead to other types of strategies, may lead to other types of interacting systems. . . . And it is very remarkable that parts of classical physics which were supposed to be in final form, or a nearly final form, are precisely some of the fields which are changing so much in the present evolution of science.[10]

Jennifer Wells, in her 2013 book *Complexity and Sustainability*, develops an overarching framework for understanding complex systems in the physical, biological, and social sciences. In complexity, each of the binaries which reigned supreme in mechanism becomes interactive: order and disorder, linear and nonlinear, predictability and unpredictability. In the physical sciences, complex dynamic systems apply to stars, suns, whirlpools, and subatomic particles. In the biological sciences, they include tree stands, ant colonies, brains, and cells. In the social sciences, complex dynamics pertain to families, cities, societies, and ideas. The principles of complexity found throughout the literature include nonlinearity, feedbacks, networks, hierarchy, emergence, and self-organization.[11]

Wells writes:

> Complexity has long been eschewed in pursuit of certainty, order, and control. Today, complexity is emerging onto the scene in all directions. Scientific and scholarly discoveries and analyses have been cracking open and exposing the dangerous simplicities of classical thinking. Finally, the dam of modernity has been broken. Meanwhile, the rising specter of great global turmoil and various threads of apocalyptic discourse are rendering more adequate conceptual tools increasingly urgent. Complexity theories are inexorably seeping through the crevices into every realm of knowledge.[12]

While most events in day-to-day life can be lived within and explained by mechanistic principles without a thought to complexity, increasingly, in cases such as climate change and sustainable ecosystems, it is necessary to understand, act upon, and live within complexity. Humanity not only has power over the nonhuman world and the capacity to alter and destroy it permanently, it also has the capacity to learn, adapt, and bring about major social and environmental changes that can alter current directions and save both the human and nonhuman worlds from deteriorating and collapsing.[13]

Theories of chaos and complexity constitute a new paradigm that goes beyond science to encompass the human relationship to nature. This new paradigm displaces the older mechanistic goals of controlling nature from outside in favor of interactive, complex systems in which human organizations are often an integral part. While chaos theories challenge the effectiveness of predictability, complexity theories argue that both nature and humanity are mutually interactive. In order to live within the new chaos and complexity paradigm, I suggest that a new ethic is critical to the human future, an approach I call partnership ethics (see below). Nonhuman nature as an autonomous organization of ecological relationships affects human social, economic, and political systems. Likewise human systems impact the functioning and indeed the very existence of the organic and inorganic elements of natural systems. Understanding nature as a complex system that includes humanity within it allows for the possibility that both *the earth as we know it today* and humanity can survive and thrive together in the coming decades.

The new chaos and complexity paradigm entails new meanings of nature that have emerged in the late twentieth and twenty-first centuries. A number of works discuss those meanings and the problems associated with the word "nature" in a post-mechanistic, complex world that is both controllable and uncontrollable, predictable and unpredictable, lawful and unruly.

Rambunctious Nature

Emma Marris characterizes the difficulties of saving a "rambunctious" nature in a post-wild world (*Rambunctious Garden*, 2011).[14] Because there is no spot on

154   *Epilogue*

earth that has not been influenced by human presence, we need to see and experience nature on the earth in a new way. "We must temper our romantic notion of untrammeled wilderness and find room next to it for the more nuanced notion of a global, half-wild rambunctious garden, tended by us."[15]

The new rambunctious garden exists in every spot on earth, including backyards, rest areas, parks, and shopping malls. The rambunctious garden does not wall off nature, but creates new nature. Evolution as an active process continues everywhere, despite the fact that the species involved may not be native to the place in which they are currently found. In fact, native ecosystems no longer exist; pristine wilderness is itself a cultural construct. While species extinctions can still be prevented, recreating the ecosystem that existed at any particular time is not achievable. The old idea of the wild must give way to the new idea of a post-wild world. That new nature is filled with human presence and the nature that is saved is one infused with human history. The new nature is an ever-changing rambunctious garden.[16]

Marris's idea of a rambunctious garden, however, has deeper historical roots and implications that go far beyond post-wild ecology. The question of a rambunctious, rebellious, and recalcitrant cosmic nature challenged philosophers from ancient times to the early modern world. Plotinus's rebellious audacity of the World Soul, the willful *natura naturans* of the Renaissance, and Francis Bacon's nature as both free and in error are examples of a self-creating nature independent of an overarching, unchanging Platonic Form or the restraint of a lawful God. Nature as chaotic, complex, and unpredictable becomes the usual mode of nature's activity, not the mechanistic paradigm that describes a more limited domain in which nature can be predicted, controlled, and managed as a resource for humanity. Today, living within an autonomous nature (as experienced on earth) entails, first and foremost, an understanding of the implications of climate change for the human and nonhuman future.

Climate Change and Autonomous Nature

In 1989, Bill McKibben in *The End of Nature* argued that there was no place on the planet that was untouched by human activity.[17] Today the effects of climate change are nowhere more evident than in the warming ice sheets in the interior of Greenland. As a result of global warming, melting glaciers are causing underlying ice sheets to crack and break free, debauching large sections of ice into the Arctic Ocean. Moreover, the rounded, melting ice crystals reflect less light back into the atmosphere, decreasing the ice fields' albedo, in turn increasing the rate of melting. The soot from wildfires in lower latitudes accelerate the decline in albedo, compounding the problem and further shrinking the ice sheets. "Amid . . . the drama of drought, fire and record heat," McKibben warns, "the planet's destiny may have been revealed . . . by the quiet metamorphosis of a silent, empty sheet of ice."[18]

In the late eighteenth century a major shift to the use of fossil fuels followed James Watt's invention of the steam engine (1784), initiating an era now referred to as the Anthropocene (subsequent to the Pleistocene, or the last ice age, and the warmer Holocene eras), characterized by the anthropogenically caused buildup of CO_2 in the atmosphere, leading to the world-wide mixing of previously isolated species and the erosion of biodiversity. The complexities and consequences of changes in interlinked climate systems, ecosystems, and human systems are extremely difficult to predict. Here, the comforts of mechanistic science have been superseded by the uncertainties of chaos and complexity theories (see above). Among the anticipated outcomes of climate change are a sea-level rise that will make many coastal and island areas uninhabitable and droughts that will cause crop failures, famines, and desertification. Some crops may benefit from increased CO_2 fertilization, while others will fail from lack of water. Many species will either migrate north or face extinction. Efforts to predict and control such events and their impacts on human life are problems discussed by the new sciences of chaos and complexity.[19]

McKibben's environmental organization, "350.org," maintains that 350 ppm (parts per million) of atmospheric carbon dioxide (CO_2) is the safe upper limit that would avoid a tipping point in life on the planet as we know it today. That number is based on NASA scientist James Hansen's contention in 2007 that

> if humanity wishes to preserve a planet similar to that on which civilization developed and to which life on Earth is adapted, paleoclimate evidence and ongoing climate change suggest that CO_2 will need to be reduced from its current 385 ppm to at most 350 ppm, but likely less than that.

Yet CO_2 reached 400 ppm in the northern industrialized countries in 2015. Moreover, the global temperature in 2015 was on track for the hottest on record.[20]

McKibben and his organization have carried these arguments around the globe in protests over global warming. Countering the continued extraction and refining of oil, including the hydraulic fracturing of shale to produce oil and natural gas that would increase atmospheric CO_2, is an ongoing endeavor. A major focus has been to defeat the Keystone Pipeline that would carry tar sands oil from Alberta, Canada, to refineries on the Gulf Coast of Texas. Arrests outside the White House in Washington have made the issue more visible. "Even if the oil manages to get safely to the refineries in Texas," McKibben asserts, "it will take a series of local problems and turn them into a planetary one . . . because those tar sands are the second-biggest pool of carbon on earth, after the oil fields of Saudi Arabia." The cumulative effects of CO_2 buildup can trigger tipping points with unpredictable cascading effects that have consequences for the human future.[21]

A World Without Humans

The dilemma that autonomous nature might ultimately achieve the upper hand over humanity has led some scholars to describe what the earth might look like if an uncontrollable, unpredictable nature wins out. Alan Weisman writes graphically about nature in a future world without humans (*The World Without Us*, 2007).[22] Climate change, environmental abuse, lack of water, poisoned soils, loss of

156   *Epilogue*

species—even without nuclear annihilation—could someday result in the loss of humans altogether. If humans disappeared, how long might the effects of their presence and artifacts last? A world from which humans have suddenly vanished would be instantly taken over by ecological and geological processes that would shape its future. "Wipe us out," he posits, "and see what's left." "How would the rest of nature respond if it were suddenly relieved of the relentless pressures we heap on it and our fellow organisms?" What would happen to farms, forests, and towns? More graphically, what would happen to architectural structures that seem so durable?[23]

The answer, Weisman says, is that changes in climate introduced by humans would continue for centuries. Carbon transferred from subterranean fossils to atmospheric gas would affect land and water alike. Saltier seas overflowing land would preserve wooden structures, wetter mountain weather would overwhelm dams, sending floods and silt to bury farms and towns below, and desert cities that once thrived on diverted water would bake and crumble. In cities, nature's processes of erosion and rot would soon triumph over human artifacts.[24]

Weisman vividly describes New York City in the days, weeks, and years following the hypothetical removal of all human presence. With the first storms, sewers would clog with debris, subways would flood, ocean waves would shut down tunnels, water mains would burst, and electrical cutoffs would cause pumps to fail. Underground soils would wash away, streets would crumble, subways would collapse, rivers would cascade through the city. Raging fires would demolish trees, wooden structures, and debris. Brick buildings would start to fall, iron structures to corrode, and steel foundations to weaken. Weeds and pioneer species would take over the landscape. Dust and soot would cover the remains. Such descriptions make the long-term consequences of climate change very graphic, leading to the possibility that *the planet as we know it today* could be radically different.[25]

Poet Gary Snyder explores the implications of "no nature" in his 1992 collection of poems entitled *No Nature*.[26] "No nature" expresses Snyder's anxiety about potential threats to planetary life from nuclear reactors, strip mining, and nondegradable plastics, aluminum cans, and junk. It poignantly exposes the dead fish and lifeless streams resulting from ecological damage. And it laments the loss of freedom of the air between us and the stars that has been taken over by jets, rockets, satellites, and war. "The greatest respect we can pay to nature," he writes, "is not to trap it, but to acknowledge that it eludes us and that our own nature is also fluid, open, and conditional." The voices that Snyder reveals in his poetry and prose are those of redwood forests, native and Asian peoples with whom he has shared much of his life, and all of earth's creatures. But nature also includes the physical earth and its industrialized and toxic spaces. There is no single concept of nature; it embraces everything that is fluid, changing, and mysterious. Ultimately, however, to "know nature" on earth is to live within it and to revere it in every way.[27]

In a post-human world, *natura naturans* transforms *natura naturata*. Nature— the creating force that brought about the world civilized by humans—becomes the chaotic, uncontrollable force that dismantles it. The limited, unusual domain of mechanistic science that allowed humans to construct and colonize the world of today gives way to the new paradigm of chaos and complexity that now describes nature's usual domain. New concepts of nature are called for that are consistent with the new paradigm.

New Concepts of Nature

Mechanistic science depends on a concept of humanity as active and nature as the passive recipient of human decisions and activities. Cartesian dualism separated mind from body, internal from external, subject from object, and science from society. A number of philosophers have attempted to redefine nature in relationship to human culture and human society. These post-modern discussions include works by Bruno Latour on natures-cultures (1993), Kate Soper on nature and the nonhuman (1995), Katherine Hayles on nature and the posthuman (1999), and Lynn Worsham and Gary Olson (2008) on the technological boundaries between humans and nature in a posthuman age. These analyses all relate to the concept of autonomous nature at the root of the new paradigm. What are these proposals and what contributions do they offer?

Natures-Cultures

Bruno Latour has elaborated on the concept of natures-cultures in *We Have Never Been Modern* (1993).[28] He addresses the problem of the dominance of modern Western culture over pre-modern and non-Western cultures by appealing to symmetry. While traditional anthropology used the assumptions of the Western world to study and assess non-Western cultures, with increasing self-reflection it became critical to change the subject-object, self-other, we-they approach to a more relativist approach. Moreover, the West tended to see nature as an object to be studied through science and to be mobilized through technology, whereas other cultures were assumed to see nature in terms of symbols and representations. As a result, instead of studying both science and nature in comparative terms, anthropologists primarily studied comparative cultures. With the rise of the sociology of knowledge, however, Western science has itself been subjected to sociological and anthropological study (witness Latour's own *Laboratory Life*, written from the viewpoint of the anthropologist in the laboratory). But nature, says Latour, should also be studied in more relativistic terms. Both Western and non-Western cultures should be viewed more symmetrically in their cultural assumptions about nature. The result is the study and comparison of what he calls natures-cultures.[29]

As Latour puts it:

> [T]he very notion of culture is an artifact created by bracketing Nature off. Cultures—different or universal—do not exist any more than Nature does. There are only natures-cultures, and these offer the only possible basis for comparison.[30]

For Latour, there is no universally agreed-upon concept of nature any more than there is a universal culture. Nature exists in the context of any given culture

158 *Epilogue*

or collective of cultures. Yet Latour does not embrace the absolute relativism that this symmetry implies. Instead of a cultural relativism imbedded in particular natures-cultures, Latour proposes a skein of networks that extend from the local to the global. These networks are built on relations and processes, not on essences and objects. They form a new world that is neither pre-modern nor anti-modern, neither modern or even post-modern, but in fact non-modern.[31]

What is non-modern is no longer separated from a nature onto which categories are imposed. Sciences and societies are co-produced and co-exist. Things and signs are inseparable. Humans are mediators at the intersections between society and nature within successive states of societies-natures. Human and nonhuman quasi-objects are hybrids. Exchanges, alliances, morphisms, and multiple forms flow together.[32]

Latour posits a new Parliament of Things in which mediators link together societies-natures. "Natures are present, but with their representatives, scientists who speak in their name. Societies are present, but with the objects that have been serving as their ballast from time immemorial." The new Parliament creates quasi-objects in an "object-discourse-nature-society whose new properties astound us all and whose network extends from my refrigerator to the Antarctic by way of chemistry, law, the State, the economy, and satellites." The Parliament includes and mediates between science and nature. "Half of our politics is constructed in science and technology. The other half of Nature is constructed in societies. Let us patch the two back together, and the political task can begin again."[33]

The outcome is an interactive system of science, technology, politics, and nature mediated through hybrids, quasi-objects, nonhuman witnesses, signs, representations, and multiple forms, i.e., natures-cultures and societies-natures. Latour's theory of natures-cultures describes the observational and laboratory aspects of science with great acuity. It characterizes a dynamic, process-oriented approach to science grounded in networks of objects and actors whose work must be seen as coupled with social-political frameworks. Latour's concept of natures-cultures as an interactive human/nature system is critical to a new paradigm for the twenty-first century and his Parliament of Things to a new ethic of partnership with nature.

What Latour's approach, needs, however, in addition to experimentation, is the other prong of "modern" science—the role of mathematics. Latour treats science itself as experiment. Nowhere does he analyze or include mathematics in his argument. While experimentation fits well into his discussion of natures-cultures and the blurring of the pre-modern and modern into skeins of networks, the networks he describes are not made of digits. Numbers in contrast to cultures and natures are in fact universals. They do not change across cultures or natures. When they move through networks they shed their cultural baggage. What distinguished modern science in the seventeenth century, in addition to advances in experimentation, was its foundation in Euclidian and analytic geometry and its invention of the calculus. Laws of nature, such as Galileo's laws of falling bodies, Boyle's law of atmospheric pressure, and Newton's laws of motion and gravitation gain

Rambunctious Nature 159

their universality from shedding their material baggage. How laws are discovered, experimentally demonstrated, and culturally debated are indeed highly significant as excursions into the history and sociology of science, but their validities remain independent of natures-cultures. Digital networks move across all cultures and across all natures.

Latour's analysis, however, is highly relevant to a new, more open, interactive, and flexible concept of nature needed for the twenty-first century. To this account, Kate Soper adds insights based on the nonhuman world in ecology and green science.

Nature and the Nonhuman World

Kate Soper in *What Is Nature? Culture, Politics, and the Non-Human* (1995) approaches the problem of nature from the perspective of discourses about nature and concerns over the meaning of nature and the non-human.[34] Her question is: Is there such a thing as nonhuman nature? Nature as "other" encompasses everything that is not human. As such it is that which humans had no role in creating. (In this sense it is akin to *natura creans/natura naturans*—nature creating/nature naturing.) Once we have mixed our labor with nature, however, it ceases to be "natural" and instead becomes cultural. Nature as nonhuman disappears. Indeed, as some would argue, nature untouched by humanity no longer exists anywhere on the planet.[35]

But underlying any discourse about nature is the humanly produced assumption of an antithesis between humanity and nature. We use the term "human" as we project our idea of the nonhuman onto nature. Yet we can also see ourselves as within a larger whole that is nature. "'Nature' in this sense is both that which we are not *and* that which we are within." The idea of humans within nature has roots in the medieval worldview and the Great Chain of Being. Here humans occupied an intermediate place between the earth and the four elements below them and the angels and God above them. Their intelligence and souls linked them to the angels and God, allowing them to understand and worship that which was above. Today ecological conceptions likewise place humanity within nature, while also seeing humans as outside it, with the technological power to fundamentally alter or destroy it.[36]

Soper sees responsibility for nature's future as a major goal of her work. In a final chapter on "Ecology, Nature, and Responsibility," she evinces concern that nature as an ecological system might be definitively altered through human activities. Here the cultural construction of nature and nature's independent existence come into tension, inasmuch as the results of pollution and climate change exist independently of human discourses about nature. Her vision for the future is one of a "genuinely democratic socialist order" and alternative "conceptions of consumption and human welfare from those promoted under capitalism. . . . " Achieving these goals requires not more "green religion," however, but more "green science." "Rather than becoming more awe-struck by nature, we need perhaps to become more stricken by the ways in which our dependency upon its resources involves

160　*Epilogue*

us irremediably in certain forms of detachment from it." Getting closer to nature, therefore, entails an understanding of how we as humans both identify with it and are different from it.[37]

Soper's work is especially valuable for its analysis of climate change and environmental pollution and for its vision of a democratic, socialist alternative to capitalist consumption and resource depletion. Yet in falling back on green science and ecology as the solution to the problems of nature and culture, she does not deal with the chaotic unpredictable dimensions of ecological change. What is also needed is an analysis that stems from chaos and complexity theories as a new ecological paradigm for the twenty-first century. Such an approach is proposed by Katherine Hayles.

Posthuman Nature

In *How We Became Posthuman* (1999), Katherine Hayles asks what the new meaning of "human" will be in a world dominated by cybernetics and described by chaos and complexity theories based on patterns and randomness rather than on presence and absence. In the Platonic and Christian worlds, meaning was derived from the presence of unchanging forms—the logos, the mind, and God—that characterized and imposed reality and meaning on a changing world. A stable, human or superhuman intelligence could identify what was real and unchanging in the messy world of appearances.[38]

Complex systems instead imply a movement toward a future that is contingent and unpredictable rather than the stable, predictable world described by mechanistic science that gave humanity a false sense of control and mastery. Randomness rather than stability becomes the ground for new creative patterns. For Hayles, the posthuman means a new dynamic relationship between intelligent, embodied humans and intelligent, embodied machines as opposed to the mechanistic ideal of domination and control by a human subject over a material world.[39]

"Cybernetics," Hayles writes, "was born in a froth of noise when Norbert Weiner first thought of it as a way to maximize human potential in a world that is in essence chaotic and unpredictable." The cybernetic noise that disrupts predictability exists both within and outside the human body and defines a changing human presence integrated within a changing world.

> The chaotic unpredictable nature of complex dynamics implies that subjectivity is emergent rather than given, distributed rather than located solely in consciousness, emerging from and integrated into a chaotic world rather than occupying a position of mastery and control removed from it.

It is in this sense that we have always been posthuman.

> Although some current versions of the posthuman point toward the antihuman and the apocalyptic, we can craft others that will be conducive to the long-range survival of humans and of the other life-forms, biological and artificial, with whom we share the planet and ourselves.

Hayles's analysis of the random nature of complex dynamics is highly relevant to the problem of living within autonomous nature that is central to the twenty-first century.[40]

Lynn Worsham and Gary Olson discuss the problems posed by new technologies for the posthuman era in their edited book, *Plugged In: Technology, Rhetoric, and Culture in a Posthuman Age* (2008).[41] How should people live in a posthuman world that blurs the boundaries between humans and the technologies they use? Our bodies, our cyberbodies, become cyborgs in which digital technologies are imbedded into human organs and limbs in ways that regulate choices and movements. Such digitally regulated human bodies are no longer human, but posthuman—or more-than-human. Human bodies and computers, whether inside the body or linked to it through fingers, heartbeats, and eyeglasses, become complex, co-adaptive systems.

Worsham and Olson's concept of a posthuman age contains implications for nature in light of the chaos and complexity paradigm. Nature becomes postnature in ways that so thoroughly blur any human/nature differences as to make a single interactive, mutually influential, and mutually interdependent post-human-nature. Nature creating (*Natura naturans*) and nature created (*Natura naturata*) merge so completely as to be inseparable and unidentifiable as a duality and instead become an interacting monism. As in the pantheistic monism of Spinoza, *natura naturans* and *natura naturata* are unified as an interacting, self-creating entity.

Autonomous Nature

The above depictions of natures-cultures, nonhuman nature, and posthuman nature all contain insights into a new relationship between humanity and nature based on the idea of autonomous nature. Autonomous nature stems from *natura creans/naturans* as the self-acting, self-creating nature of past philosophies in which nature is both creating and rebellious, predictable and unpredictable, lawful and unlawful. Autonomous nature is the nature at the root of chaos theory. And it is the nature at the edge of order and chaos in complexity theory. The laws of nature defined by Newton and Leibniz make nature orderly, predictable, and manageable in the everyday world. But the chaotic nature of hurricanes, droughts, volcanic eruptions, earthquakes, and epidemics, along with uncertainty principles in quantum mechanics and nonlinear dynamics, make it unpredictable. Order and chaos, linear and nonlinear, natures-cultures, and human/nonhuman all exist and interact in the everyday world of the twenty-first century. Autonomous nature is the nature at the root of the new chaos and complexity paradigm in which humans and nonhuman nature must exist together and thrive.

Partnership Ethics

What new ethical systems are needed in a world in which the boundaries of nature and culture are blurred and in which nature is autonomous? At the cusp between human and nonhuman nature exists life and living things that feel pain and pleasure. Anthropocentric ethical systems that originated with the utilitarianism of

162   *Epilogue*

Jeremy Bentham and John Stuart Mill evolve toward systems that include animals and the living world. With the rise of the environmental justice movement in the late 1980s, ethics expands to include the rights of minorities with respect to the environment. By merging anthropocentric with ecocentric ethics—the ethic that includes all of nature within it—we can develop an integrated, interactive ethic based on partnership between the human and nonhuman worlds.

To incorporate and include concepts of earthly nature as active and often unpredictable and unmanageable in the twenty-first century, I have proposed a partnership ethic. This ethic recognizes both nature and human communities as creative, changing, and interacting entities. It states: *A partnership ethic holds that the greatest good for the human and nonhuman communities is in their mutual living interdependence.*[42]

My ethic contains five precepts for a human community in a sustainable partnership with a nonhuman community:

- Equity between the human and nonhuman communities.
- Moral consideration for both humans and other species.
- Respect for both cultural diversity and biodiversity.
- Inclusion of women, minorities, and nonhuman nature in the code of ethical accountability.
- An ecologically sound management that is consistent with the continued health of both the human and the nonhuman communities.

A partnership ethic entails a viable relationship between a human community and a nonhuman community in a particular place, a place in which connections to the larger world are recognized through economic and ecological exchanges. It is an ethic in which humans act to fulfill both humanity's vital needs and nature's needs by restraining human hubris.

A partnership ethic is a synthesis between an ecological approach (or ecocentric ethics) based on moral consideration for all living and nonliving things and a utilitarian human-centered (i.e., anthropocentric/homocentric) approach based on the social good and the fulfillment of basic human needs. All humans have needs for food, clothing, shelter, and energy, but nature also has an equal need to survive. The new ethic questions the notion of the unregulated market, sharply criticizing egocentric ethics—what is good for the individual is good for society—and instead proposes a partnership between nonhuman nature and the human community. It avoids gendering nature as a nurturing mother or a goddess and avoids the ecocentric dilemma that humans are only one of many equal parts of an ecological web and therefore morally equal to a bacterium or a mosquito.

A partnership ethic brings humans and nonhuman nature into a dynamically balanced, more nearly equal relationship with each other. Humans, as the bearers of ethics, acknowledge nonhuman nature as an autonomous actor which cannot be predicted or controlled except in very limited domains. We acknowledge that we have the potential to destroy life as we currently know it through nuclear power, pesticides, toxic chemicals, and unrestrained economic development and

Rambunctious Nature 163

act to exercise specific restraints on that ability. We cease to create profit for the few at the expense of the many. We instead organize our economic and political forces to fulfill people's vital needs for food, clothing, shelter, and energy, and to provide security for health, jobs, education, children, and old age. Such forms of security would rapidly reduce population growth rates since a major means of providing security would not depend on having large numbers of children or on economies in which boys are favored over girls as is the case in many countries today.

A partnership ethic treats humans (including male partners and female partners) as equals in personal, household, and political relations and humans as equal partners with (rather than controlled by or dominant over) nonhuman nature. Just as human partners, regardless of sex, race, or class must give each other space, time, and care, allowing each other to grow and develop individually within supportive nondominating relationships, so humans must give nonhuman nature space, time, and care, allowing it to reproduce, evolve, and respond to human actions. In practice, this would mean not cutting forests and damming rivers that make people and wildlife in flood plains more vulnerable to "natural disasters"; curtailing development in areas subject to volcanic eruptions, earthquakes, hurricanes, and tornados to allow room for unpredictable, chaotic, natural surprises; and exercising ethical restraint in introducing new technologies such as pesticides, genetically engineered organisms, and biological weapons into ecosystems. Constructing nature as a partner allows for the possibility of a personal or intimate relationship with nature and for feelings of compassion for nonhumans as well as for people who are sexually, racially, or culturally different from ourselves.

If we know that a major earthquake in Los Angeles is likely in the next seventy-five years, a utilitarian, homocentric ethic would state that the government ought not to license the construction of a nuclear reactor on the faultline. But a partnership ethic says that, we, the human community, ought to respect nature's autonomy as an actor by limiting building and leaving open space. If we know there is a possibility of a one-hundred-year flood on the Mississippi River, we respect human needs for navigation and power, but we also respect nature's autonomy by limiting our capacity to dam every tributary that feeds the river and build homes on every flood plain. We leave some rivers wild and free and leave some flood plains as wetlands, while using others to fulfill human needs. If we know that forest fires are likely in the Rockies, we do not build cities along forest edges. We limit the extent of development, leave open spaces, plant fire-resistant vegetation, and use tile rather than shake roofs. If cutting tropical and temperate old-growth forests creates problems for both the global environment and local communities, but we cannot adequately predict the outcome or effects of those changes, we need to conduct partnership negotiations in which nonhuman nature and the people involved are equally represented.

Partnership ethics is especially important to a solar-based, non-fossil-fueled world. Large solar arrays and wind generators, while critical to limiting greenhouse gases, can be detrimental to bird life. In the massive Ivanpah solar array constructed in California's Mojave Desert, birds flying too close to the panels and the towers have been killed in flight. Birds and bats, including endangered species,

164   *Epilogue*

have been decimated by wind turbines. Implementing a partnership ethic in such cases involves negotiations at all levels, including biologists, energy companies, the federal government, and local residents. Solutions include both technological and biological deterrents such as radar detection followed by recorded predator calls, trained falcons and hawks in the air, and working dogs on the ground. Here autonomous nature and human partnerships come together to promote sustainability in the more-than-human world.[43]

Partnership ethics makes visible the connections between people and the environment in an effort to find new cultural and economic forms that fulfill vital needs, provide security, and enhance the quality of life without degrading the local or global environment. It creates both a structure and a set of goals that can enable decision making, consensus, and mediation to be achieved without contentious litigation. It relates work in the sciences of ecology and chaos and complexity theories to new possibilities for nondominating relationships between humans and nonhuman nature.

Ethics and justice are essential ingredients in the potential for a productive response to climate change as the primary exemplar of autonomous nature. Every major aspect of climate negotiation is an ethical issue; ethical principles and reasoning are needed in order to work through the challenging issues on the negotiating table. Climate change likewise greatly alters the larger field of social justice. It especially impacts people of minority cultures, often requiring new theories of ethics and justice to deal with the crises created. Climate justice has different meanings for Alaskan Natives, American Indians, African Americans, and Latinos. The example of climate change as a problem of the predictability of active nature therefore incorporates the interlinked issues of morals, ethics, and justice associated with ideas of natural law and the laws of nature.[44]

Climate policy must not only incorporate environmental justice, but also sustainable energy and sustainable food systems. To paraphrase the policy of "a chicken in every pot and a car in every garage," attributed to Herbert Hoover in 1928, we might say: "Solar panels on every roof, bicycles in every garage, and vegetables in every backyard." If every building had solar panels on its roof, wind generators along its roof line, and efficient storage batteries, communities would move a long way down the road to energy self-sufficiency. If most people traveled short distances by bicycle and longer distances by solar-charged electric vehicles, greenhouse gas emissions would be greatly reduced. If vegetables were grown on rooftops, in backyards, and in solar greenhouses, many people could be nourished at lower levels of the food chain. New policies in which humanity and earthly nature are in partnership would go a long way toward making the earth of the future a sustainable place to live.

Conclusion

In this book, I have traced the evolution of the concepts and tensions between *natura naturans,* as active, creative, and potentially uncontrollable nature and *natura naturata,* as the created world that can be described by mathematics and

Rambunctious Nature 165

experimentation. I have offered a new look at the origins of early modern science in the seventeenth-century Scientific Revolution as a prehistory of chaos and complexity theories. I have argued that the concept of autonomous nature captures for the twenty-first century what those earlier terms represented for the ancient, medieval, and early modern eras. The lesson for the twenty-first century is to see how humanity can live within an autonomous earthly nature—a nature that is both predictable and unpredictable, controllable and uncontrollable, orderly and disorderly. To live in ways that allow both human and more-than-human nature to survive and thrive will take new forms of thinking, feeling, and acting. Those forms include new sciences, philosophies, politics, economics, and partnerships.

Notes

1 Thomas Merton, "Hagia Sophia," *Dawn. The Hour of Lauds. Collected Poems* (New York: New Directions, 1977), p. 361. By Thomas Merton, from THE COLLECTED POEMS OF THOMAS MERTON, copyright ©1963 by The Abbey of Gethsemani. Reprinted by permission of New Directions Publishing Corp.
2 Ed Pilkington, "The Village at the Tip of the Iceberg," in *The Guardian/The Observer*, September 28, 2008, http://www.theguardian.com/environment/2008/sep/28/alaska.climatechange.
3 Thomas Kuhn, *The Structure of Scientific Revolutions* (Chicago: University of Chicago Press, 1962), esp. pp. 10–20.
4 Kuhn, *Structure of Scientific Revolutions*, quotation on p. 91.
5 On the history of probability and statistics, see Lorenz Krüger, Lorraine Daston, and Michael Heidelburger, eds., *The Probablistic Revolution*, 2 vols. (Cambridge, MA: MIT Press, 1987); Ian Hacking, *The Emergence of Probability: A Philosophical Study of Early Ideas About Probability, Induction, and Statistical Inference* (New York: Cambridge University Press, 2006); Stephen M. Stigler, *The History of Statistics: The Measurement of Uncertainty Before 1900* (Cambridge, MA: Harvard University Press, 1986). Scholars have debated the extent to which probability theory constituted a scientific revolution versus an evolution or emergence. Of particular interest is the two-volume work, *The Probabilistic Revolution*, ed. Krüger, Daston, and Heidelberger. In the introductory essay, "What Are Scientific Revolutions?" (pp. 7–22), Thomas Kuhn states:

> It is now almost twenty years since I first distinguished what I took to be two types of scientific development, normal and revolutionary. . . . Revolutionary change is defined in part by its difference from normal change, and "normal" change is . . . the sort that results in growth, accretion, [and] cumulative addition to what was known before. . . . Revolutionary changes are different and far more problematic. They involve discoveries that cannot be accommodated within the concepts in use before they were made. In order to make or to assimilate such a discovery one must alter the way one thinks about and describes some range of natural phenomena. (pp. 7, 8)

Bernard Cohen, in "Scientific Revolutions, Revolutions in Science, and a Probabilistic Revolution, 1800–1930" (pp. 23–44), argues for a "probabilizing revolution." He writes:

> In the twentieth century, we have witnessed a true revolution in the physical sciences resulting in the substitution of probability and statistics for the old Newtonian simple causality of assigned cause and effect. A similar movement in biology has centered on genetics and evolution. There is ample testimony to this revolution, beginning in the social sciences in the mid-nineteenth century and reaching a culmination in the physics of radioactivity and quantum mechanics in the first third of the

166 *Epilogue*

twentieth century. . . . [E]ven if the decades 1800–1930 do not show a single revolution in the domain of probability, they provide evidence of a *probabilizing revolution*, that is of a true revolution of fantastic consequences attendant on the introduction of probability and statistics into areas that have undergone revolutionary changes as a result. (p. 40)

Ian Hacking interprets it as an "erosion of determinism" and a "taming of chance" (p. 54). In "Was There a Probabilistic Revolution 1800–1930?" (pp. 45–55), he discusses several changes that occurred during the period, but wonders whether the segments add up to a revolution. He concludes,

The research reported in these volumes began with a simple observation: today our vision of the world is permeated by probability, while in 1800, it was not. Probability is the great philosophical success story of the period. . . . It has been a success in metaphysics, epistemology, and pragmatics, to mention but three of the classic philosophical fields. . . . In 1800 the world was deemed to be governed by stern necessity and universal laws. Shortly after 1930 it became virtually certain that at bottom our world is run at best by laws of chance. (p. 45)

Lorenz Krüger, in "The Slow Rise of Probabilism" (pp. 59–89), argues that probability as mainstream theory arose very slowly during the nineteenth century in part because determinism, mechanism, and the desire for the domination of nature were (and are) such powerful and attractive ideas. In volume 2, numerous applications of probability are delineated in the fields of psychology, sociology, physiology, evolutionary biology, and physics.

6 Edward Lorenz, *The Essence of Chaos* (Seattle: University of Washington Press, 1993); Ilya Prigogine and Isabelle Stengers, *Order out of Chaos: Man's New Dialogue with Nature* (New York: Bantam, 1984); James Gleick, *Chaos: Making a New Science* (New York: Viking, 1987); M. Mitchell Waldrop, *Complexity: The Emerging Science at the Edge of Order and Chaos* (New York: Simon and Schuster, 1992).

7 Gleick, *Chaos: Making a New Science*, p. 6.

8 Edward Lorenz, "Predictability: Does the Flap of a Butterfly's Wings in Brazil Set off a Tornado in Texas?" in *The Essence of Chaos*, ed. Lorenz (Seattle: University of Washington Press, 1993), pp. 181–184; Quotations are from Lorenz, *Essence of Chaos*, pp. 10, 79, 13. See also Carolyn Merchant, "Introduction," *Key Concepts in Critical Theory: Ecology*, 2nd ed., ed. Merchant (Amherst, NY: Humanity Books, 2008), from p. 34.

9 Merchant, "Introduction," *Key Concepts in Critical Theory: Ecology*, from p. 34.

10 Ilya Prigogine, "The Rediscovery of Time: Science in a World of Limited Predictability," paper presented to the International Congress on Spirit and Nature, Hanover, Germany, May 21–27, 1988, reprinted in Merchant, *Key Concepts in Critical Theory: Ecology*, pp. 401–406, quotation on p. 405. See also Ilya Prigogine and Isabelle Stengers, *Order out of Chaos: Man's New Dialogue with Nature* (New York: Bantam, 1984) and Erich Jantsch, *The Self-Organizing Universe* (New York: Pergamon, 1980).

11 Jennifer Wells, *Complexity and Sustainability* (New York: Routledge, 2013), Ch. 2, "Elucidating Complexity Theories," pp. 19–51.

12 Wells, *Complexity and Sustainability*, p. 50.

13 Wells, *Complexity and Sustainability*, pp. 49–51.

14 Emma Marris, *Rambunctious Garden: Saving Nature in a Post-Wild World* (New York: Bloomsbury, 2011).

15 Marris, *Rambunctious Garden*, pp. 1–5, quotation on p. 2.

16 Marris, *Rambunctious Garden*, pp. 2–3, 12, 14–15.

17 Bill McKibben, *The End of Nature* (New York: Random House, 1989).

18 Bill McKibben, "The Arctic Ice Crisis: Greenland's Glaciers Are Melting Far Faster than Scientists Expected," *Rolling Stone*, August 30, 2012.

Rambunctious Nature   167

19 Paul J. Crutzen and Eugene F. Stoermer, "The Anthropocene," *IGPB* (*International Geosphere-Biosphere Programme*) *Newsletter* 41 (2000): 17; Paul J. Crutzen, "Geology of Mankind," *Nature* (January 3, 2002): 23; Dipesh Chakrabarty, "The Climate of History: Four Theses," *Critical Inquiry* 35, no. 2 (Winter 2009): 197–222; Jennifer Wells and Carolyn Merchant, "Melting Ice: Climate Change and the Humanities," *Confluence* XIV, no. 2 (Spring 2009): 13–27; Wells, *Complexity and Sustainability*; McKibben, *End of Nature*.

20 Bill McKibben, "Global Warming's Terrifying New Math," *Rolling Stone*, August 2, 2012. James Hansen et al., "Target Atmospheric CO_2: Where Should Humanity Aim?" *Open Atmospheric Science Journal* 2 (2008): 217–231; for data on changing CO_2 levels see, http://www.climatecentral.org/news.

21 Bill McKibben, "The Keystone Pipeline Revolt: Why Mass Arrests Are Just the Beginning," *Rolling Stone*, October 13, 2011.

22 Alan Weisman, *The World Without Us* (New York: Picador Press, 2007).

23 Weisman, *The World Without Us*, pp. 4–5, quotation on p. 5.

24 Weisman, *The World Without Us*, pp. 22–24.

25 Weisman, *The World Without Us*, pp. 26–33.

26 Gary Snyder, *No Nature: New and Selected Poems* (New York: Pantheon, 1992).

27 Snyder, *No Nature*, quotation in "Preface." For examples, see "LMFBR (Liquid Metal Fast Breeder Reactor)," p. 246; "Bomb Test," p. 325; and "Strategic Air Command," p. 279.

28 Bruno Latour, *We Have Never Been Modern* (Cambridge, MA: Harvard University Press, 1993).

29 Latour, *We Have Never Been Modern*, pp. 91–96.

> [C]omparative anthropology ... no longer compares cultures, setting aside its own, which through some astonishing privilege possesses a unique access to universal Nature. *It compares nature-cultures.* Are they comparable? Are they similar? Are they the same? We can now, perhaps solve the insoluble problem of relativism. (p. 96)

30 Latour, *We Have Never Been Modern*, p. 104.

> Still the problem of relativism has not been solved. Only the confusion resulting from the bracketing off of Nature has been provisionally eliminated. We now find ourselves confronting productions of nature-cultures that I am calling collectives—as different, it should be recalled from the society construed by sociologists—men-among-themselves—as they are from the Nature imagined by epistemologists—things in themselves. (pp. 106–107)

31 Latour, *We Have Never Been Modern*, pp. 106, 120–122, 129, 130–132.

32 Latour, *We Have Never Been Modern*, pp. 134, 137, 139.

33 Latour, *We Have Never Been Modern*, quotations on p. 144.

34 Kate Soper, *What Is Nature? Culture, Politics, and the Non-Human* (Oxford: Blackwell, 1995).

35 Soper, *What Is Nature*, pp. 15, 16, 18, 20.

36 Soper, *What Is Nature*, pp. 15, 21–23, 143–144, 249–250, quotation on p. 21.

37 Soper, *What Is Nature*, pp. 249, 277, quotations on pp. 271, 278.

38 N. Katherine Hayles, *How We Became Posthuman: Virtual Bodies in Cybernetics, Literature, and Informatics* (Chicago: University of Chicago Press, 1999), pp. 283–285.

39 Hayles, *How We Became Posthuman*, pp. 285–288.

40 Hayles, *How We Became Posthuman*, pp. 288–291, quotations on p. 291.

41 Lynn Worsham and Gary A. Olson, eds. *Plugged In: Technology, Rhetoric, and Culture in a Posthuman Age* (Cresskill, NJ: Hampton Press, 2008).

42 This section on partnership ethics draws on and is excerpted from my books: Carolyn Merchant, "Partnership Ethics," in Merchant, *Earthcare: Women and the Environment*

168   *Epilogue*

(New York: Routledge, 1996), pp. 209–224; Merchant, *Radical Ecology: The Search for a Livable World*, 2nd ed. (New York: Routledge, 2006), pp. 83–85, 197; Merchant, "Partnership," in Merchant, *Reinventing Eden: The Fate of Nature in Western Culture* (New York: Routledge, 2003, 2013), see 1st ed., pp. 223–246; 2nd ed., pp. 191–206.

43   On solar arrays and birds see, http://www.renewableenergyworld.com/rea/news/article/2014/09/preventing-bird-deaths-at-solar-power-plants-part-1. On wind turbines and birds, see http://grist.org/climate-energy/for-the-birds-and-the-bats-8-ways-wind-power-companies-are-trying-to-prevent-deadly-collisions/.

44   Wells and Merchant, "Melting Ice," 13–27.

Bibliography

Adam, Barbara. *Timescapes of Modernity: The Environment and Invisible Hazards.* New York: Routledge, 1998.

Alberti, Leon Battista. *On Painting.* Edited by John Richard Spencer. New Haven, CT: Yale University Press, 1956.

Alberti, Leon Battista. *On Painting: A New Translation and Critical Edition.* Edited and translated by Rocco Sinisgalli. New York: Cambridge University Press, 2011.

Alberti, Leon Battista. *On Painting and On Sculpture: The Latin Texts of* De Pictura *and* De Statua. Edited with translation, introduction, and notes by Cecil Grayson. London: Phaidon, 1972.

Alberti, Leon Battista. *On the Art of Building in Ten Books.* Translated by N. Leach, J. Rykwert, and R. Tavenor. Cambridge: MIT Press, 1988.

Ames, Christine Caldwell. *Righteous Persecution: Inquisition, Dominicans, and Christianity in the Middle Ages.* Philadelphia: University of Pennsylvania Press, 2009.

Anonymous. *A True and Fearefull Vexation of One Alexander Nyndge: Being Most Horribly Tormented with the Deuill, From the 20. Day of Ianuary, to the 23. of Iuly. At Lyering Well in Suffocke: with his prayer afer his deliuerance. Written by his owne brother Edvvard Nyndge Master of Arts, with the names of the witnesses that were at his vexation.* London: W. Stansby[?], 1615.

Aquinas, Thomas. *Selected Writings.* Edited and translated by Ralph McInerny. New York: Penguin, 1998.

Aquinas, Thomas. *Summa Theologica.* Latin text and English translation. London: Blackfriars, 1963; Vol. 26, "Original Sin," Question 85, "Sin's Damage to the Good of Nature," Article 6, "Whether or not death and similar evils are natural to man."

Aristotle. *The Basic Works of Aristotle.* Edited by Richard McKeon. New York: Random House, 1941.

"Augustine of Hippo." http://en.wikipedia.org/wiki/Augustine_of_Hippo.

Augustine, Saint. *The City of God (De Civitate Dei).* Translated by John Healy. London: J.M. Dent, 1903.

Augustine, Saint. *Confessions.* http://sparks.eserver.org/books/augustineconfess.pdf.

Augustine, Saint. *Confessions.* Translated by Henry Chadwick. Oxford, UK: Oxford University Press, 1991.

Augustine, Saint. *The Confessions of Saint Augustine, A Text and Commentary.* Edited by J.J. O'Donnell. Oxford, UK: Oxford University Press, 1992. http://www.stoa.org/hippo/noframe_entry.html.

Augustine, Saint. *On the Trinity [De Trinitate (ca. 428 C.E.)], Books 8–15.* Edited by Gareth B. Matthews. Translated by Stephen McKenna. Cambridge, UK: Cambridge University Press, 2002.

170 Bibliography

Augustine, Saint. *De Trinitate* (Latin). http://www.thelatinlibrary.com/august.html.

Bacon, Francis. *Novum Organum*. Edited by Thomas Fowler. London: Clarendon Press, 1878; 2nd ed., 1889.

Bacon, Francis. *Works*. Edited by James B. Spedding, Robert Leslie Ellis, and Douglas Devon Heath. 14 vols. London: Longmans Green, 1868–1901. *Wisdom of the Ancients* (1609); *Thema Coeli* (1612); *Parasceve* (1620); *The Great Instauration* (1620); *Novum Organum* (1620); *Advancement of Learning* (1605, 1623); *New Atlantis* (1627).

Bacon, Francis. *Works* (Latin). Edited by James B. Spedding, Robert Leslie Ellis, and Douglas Devon Heath. 14 vols. London: Longmans Green, 1857–1874, vol. 1. *De Dignitate et Augmentis Scientiarum* (1623).

Ball, Philip. "Alchemical Culture and Poetry in Early Modern England." *Interdisciplinary Reviews* 31, no. 1 (2006), p. 12.

Battisti, Eugenio. "Natura Artificiosa to Natura Artificialis." In *The Italian Garden*. Washington, DC: Dumbarton Oaks and the Trustees for Harvard University, 1972.

Bett, Henry. *Johannes Scotus Erigena: A Study in Medieval Philosophy*. Cambridge, UK: Cambridge University Press, 1925.

Bialostocki, Jan. "The Renaissance Concept of Nature and Antiquity." In *The Renaissance and Mannerism: Studies in Western Art: Acts of the Twentieth International Congress of the History of Art*, vol. 2, edited by Millard Meiss et al. Princeton, NJ: Princeton University Press, 1963, pp. 19–30.

Blumenberg, Hans. "Toward a Prehistory of the Idea of the Creative Being." Translated by Anna Wertz. *Que Parle*. Special issue on *The End of Nature* 12, no. 1 (Spring/Summer 2000): 17–54.

Boccaccio, Giovanni. *The Decameron*. Translated by Richard Aldington. Garden City, NY: International Collectors Library, 1930.

Bogaert, Pierre-Maurice. "The Latin Bible, c. 600 to c. 900." In *From 600 to 1450. The New Cambridge History of the Bible*, vol. 2, edited by Richard Marsden and E. Ann Matter. Cambridge, UK: Cambridge University Press, 2012, pp. 69–92.

Bonaventure, Saint. *Opera omnia . . . edita studio et cura PP Collegii a S. Bonaventura, ad plurimos codices mss. emendata, anecdotis aucta, prolegomenis scholiis notisque illustrata*. 11 vols. Ad Claras Aquas, Quarracchi, Ex typographia Collegii S. Bonaventurae, 1882–1902.

Bonaventure, Saint. *The Works of Bonaventure*. Translated by Jose de Vinck. 4 vols. Patterson, NJ: St. Anthony Press, 1963.

Botkin, Daniel. *Discordant Harmonies: A New Ecology for the Twenty-First Century*. New York: Oxford University Press, 1990.

Botkin, Daniel. *The Moon in the Nautilus Shell: Discordant Harmonies Reconsidered, From Climate Change to Species Extinction, How Life Persists in an Ever-Changing World*. New York: Oxford University Press, 2012.

Bowerbank, Sylvia Lorraine. *Speaking for Nature: Women and Ecologies of Early Modern England*. Baltimore and London: John Hopkins University Press, 2004.

Boyle, Robert. *General History of the Air*. London: Awnsham and John Churchill, 1692.

Braun, Theodore E.D. and John B. Radner. *The Lisbon Earthquake: Representations and Reactions*. Oxford, UK: Voltaire Foundation, 2005.

Brett, George Sidney. *The Philosophy of Gassendi*. London: MacMillan, 1908.

Brose, Eric Dorn. *Technology and Science in the Industrializing Nations, 1500–1914*. Amherst NY: Humanity Books, 2006.

Bruno, Giordano. *Cause, Principle, and Unity*. Translated by Jack Lindsay. New York: International Publishers, 1962.

Bibliography   171

Bruno, Giordano. *Cause, Principle, and Unity and Essays on Magic*. Edited by Richard J. Blackwell and Robert de Lucca with an Introduction by Alfonso Ingegno. Cambridge, UK: Cambridge University Press, 1998.

Bruno, Giordano. *De Monade, numero et figura*. In *Opera Latine Conscripta*. Naples: Morano, 1884, vol. I, Pt. II.

Caird, John. *Spinoza*. Edinburgh: Blackwood, 1888.

Campbell, Bruce M.S., ed. *Before the Black Death: Studies in the "Crisis" of the Early Fourteenth Century*. Manchester, UK: Manchester University Press, 1991.

Cañizares-Esguerra, Jorge. *Nature, Empire, and Nation: Explorations of the History of Science in the Iberian World*. Stanford, CA: Stanford University Press, 2006.

Cannon, Terry. "Vulnerability, 'Innocent' Disasters, and the Imperative of Cultural Understanding." *Disaster Prevention and Management* 17, no. 3 (2008): 350–357.

Chakrabarty, Dipesh. "The Climate of History: Four Theses." *Critical Inquiry* 35, no. 2 (Winter 2009): 197–222.

Chamber's Encyclopedia: A Dictionary of Universal Knowledge. Philadelphia: Lippincott, 1889. Vol. 4, "Erigena," p. 413.

Chester, David. "The Theodicy of Natural Disasters." *Scottish Journal of Theology* 51, no. 4 (1998): 485–505.

Chester, David K. and A.M. Duncan. "Volcanoes, Earthquakes, and God: Christian Perspectives on Natural Disasters." *SECED Newsletter* (The Society for Earthquake and Civil Engineering Dynamics) 22, no. 4 (2011): 1–6.

Cicero. *Laws [De Legibus]*. With an English translation by C.W. Keyes. Loeb Classical Library. Cambridge, MA: Harvard University Press, 1970, vol. 16, pp. 297–369.

Cicero. *De Legibus* (Latin). http://www.thelatinlibrary.com/cicero/leg1.shtml.

Cipolla, Carlo M. *Before the Industrial Revolution: European Society and Economy, 1000–1700*. New York: W.W. Norton, 1976.

Clark, Robert David. "*Natura creatrix*: The Matter of Meaning in the *De Rerum Natura*." PhD dissertation, Columbia University, 2000.

Coffin, David R. "The Study of the History of the Italian Garden until the First Dumbarton Oaks Colloquium." In *Perspectives on Garden History*, edited by Michel Conan. Washington, DC: Dumbarton Oaks Research Library, 1999, pp. 27–35.

Cole, Rosalie L. "Spinoza and the Early English Deists." *Journal of the History of Ideas* 20, no. 1 (January 1959): 23–46.

Colerus, Jean. *La Vie de B. de* Spinoza (1706). In *Vies de Spinoza*. Paris: Allia, 1999, pp. 7–91.

Collingwood, R.G. *The Idea of Nature*. Oxford, UK: Oxford University Press, 1939.

Conan, Michel, ed. *Perspectives on Garden History*. Washington, DC: Dumbarton Oaks Research Library, 1999.

Cronon, William, ed. *Uncommon Ground: Rethinking the Human Place in Nature*. New York: W.W. Norton, 1995; 2nd ed. 1996.

Crutzen, Paul J. "Geology of Mankind." *Nature* (January 3, 2002): 23.

Crutzen, Paul J. and Eugene F. Stoermer, "The Anthropocene." *IGPB* (*International Geosphere-Biosphere Programme*) *Newsletter* 41 (2000): 17.

Cusa, Nicholas of. *Idiota de mente* (*The Layman, About Mind*). Translated with an Introduction by Clyde Lee Miller. New York: Abaris Books, 1979.

Cusa, Nicholas of. *On Learned Ignorance* (*De Docta Ignoratia*). Translated by Jasper Hopkins. Minneapolis, MN: A.J. Benning, 1981.

Cusa, Nicholas of. *Selections, English and Latin*. Cambridge, MA: Harvard University Press, 2008.

172   Bibliography

Darrel, John. *A True Narration of the Strange and Grevous Vexation by the Devil, of 7 Persons in Lancashire, and William Somers of Nottingham* (England [?]: English Secret Press [?], 1600.

Daston, Lorraine and Michael Stolleis, eds. *Natural Law and Laws of Nature in Early Modern Europe*. Burlington, VT: Ashgate, 2008.

Davis, Edward. "Newton's Rejection of the 'Newtonian World View': The Role of Divine Will in Newton's Natural Philosophy." *Science and Christian Belief* 3, no. 1 (1991): 103–117.

Deck, John. *Nature, Contemplation, and the One: A Study in the Philosophy of Plotinus*. Toronto: University of Toronto Press, 1967.

De Dijn, Hermann. "The Articulation of Nature, or the Relation God-Modes in Spinoza." *Giornale Critico della Filosofia Italiana* 8, nos. 3–4 (1977): 337–344.

Deely, John. *Four Ages of Understanding: The First Postmodern Survey of Philosophy from Ancient Times to the Turn of the Twenty-First Century*. Toronto: University of Toronto Press, 2001.

De Jonge, Eccy. *Spinoza and Deep Ecology: Challenging Traditional Approaches to Environmentalism*. Burlington, VT: Ashgate, 2004.

Descartes, René. In *The Philosophical Works of Descartes*. Edited by E. S. Haldane and G.R.T. Ross. 2 vols. New York: Dover, 1955.

Descartes, René. *Le Monde, ou Traite de la lumiere*. Translated with an Introduction by Michael Sean Mahoney. New York: Abaris Books, 1979.

Descartes, René. *Passions of the Soul*. Translated and annotated by Stephen Voss. Indianapolis, IN: Hackett, 1989.

Descartes, René. *The World and Other Writings*. Edited by Stephen Gaukroger. Cambridge, UK: Cambridge University Press, 1998.

Descola, Philippe. *The Ecology of Others*. Translated by Genevieve Godbout and Benjamin P. Luley. Chicago: Prickly Paradigm Press, 2013.

Devall, Bill. *Simple in Ends, Rich in Means: Practicing Deep Ecology*. Salt Lake City, UT: Peregrine Smith Books, 1988.

Devall, Bill and George Sessions, eds. *Deep Ecology: Living as if Nature Mattered*. Salt Lake City, UT: G.M. Smith, 1985.

Dobbs, Betty Jo Teeter. *The Foundations of Newton's Alchemy: or, "The Hunting of the Greene Lyon."* New York: Cambridge University Press, 1975.

Dobbs, Betty Jo Teeter. *The Janus Faces of Genius: The Role of Alchemy in Newton's Thought*. New York: Cambridge University Press, 1991.

Dobbs, Betty Jo Teeter and Margaret C. Jacob. *Newton and the Culture of Newtonianism*. Atlantic Highlands, NJ: Humanities Press, 1995.

Doueihi, Milad. *Augustine and Spinoza*. Translated by Jane Marie Todd. New York: Oxford University Press, 2010.

Dronke, Peter. "Bernard Silvestris, Natura, and Personification." *Journal of the Warburg and Courtauld Institutes* 43 (1980): 16–31.

Dynes, Russell R. "The Dialogue Between Voltaire and Rousseau on the Lisbon Earthquake: The Emergence of a Social Science View." *International Journal of Mass Emergencies and Disasters* 18, no. 1 (March 2000): 97–115.

Economou, George. *The Goddess Natura in Medieval Literature*. Cambridge, MA: Harvard University Press, 2002 [1972].

Epicurus. *The Extant Remains*. Translated by Cyril Bailey. Oxford, UK: Clarendon Press, 1926.

Erigena, Johannes Scotus. *Periphyseon: Joannis Scoti Erigenae de divisione naturae*. Oxford, 1681. [ed. Thomas Gale, Wing J 747].

Eriugena, Joannes Scotus. *Periphyseon: On the Division of Nature.* Translated by Myra L. Uhlfelder with summaries by Jean A. Potter. Indianapolis, IN: Bobbs-Merrill, 1976.

Eriugena, Johannis Scotti. *Periphyseon* (*De Divisione Naturae*), *English and Latin.* Edited by I. P. Sheldon-Williams, with the collaboration of Ludwig Bieler. Dublin: Institute for Advanced Studies, 1968–.

Fowler, Don. *Lucretius on Atomic Motion: A Commentary on* De Rerum Natura, *Book Two, lines 1–332.* Oxford, UK: Oxford University Press, 2002.

Fuller, B[enjamin] A[pthorp] G[ould]. *A History of Philosophy.* Edited by Sterling M. McMurrin. New York: Holt, 1960 [1955].

Gale, Monica R. *Virgil on the Nature of Things: The Georgics, Lucretius, and the Didactic Tradition.* Cambridge, UK: Cambridge University Press, 2000.

Garber, Daniel. *Descartes Embodied: Reading Cartesian Philosophy Through Cartesian Science.* New York: Cambridge University Press, 2001.

Garber, Daniel. *Descartes' Metaphysical Physics.* Chicago: University of Chicago Press, 1992.

Garber, Daniel. *Leibniz: Body, Substance, Monad.* New York: Oxford University Press, 2009.

Gardner, Alice. *Studies in John the Scot (Erigena): A Philosopher of the Dark Ages.* London: Henry Frowde, 1900.

Garrard, Mary D. "Leonardo da Vinci: Female Portraits, Female Nature." In *The Expanding Discourse: Feminism and Art History*, edited by Norma Broude and Mary D. Garrard. New York: HarperCollins, 1992, pp. 58–86.

Gassendi, Pierre. *The Selected Works of Gassendi.* Translated by Craig B. Bush. London: Johnson Reprint, 1972.

Gassendi, Pierre. *Syntagma philosophicum* in *Opera Omnia.* 6 vols. Stuttgart-Bad Cannstatt: Froman, 1954 [1658], facsimile edition.

Glacken, Clarence. *Traces on the Rhodian Shore: Nature and Culture in Western Thought from Ancient Times to the End of the Eighteenth Century.* Berkeley: University of California Press, 1967.

Gladhill, Charles William. "*Foedera*: A Study in Roman Poetics and Society." PhD dissertation, Stanford University, 2008.

Gleick, James. *Chaos: Making a New Science.* New York: Viking, 1987.

Golinski, Jan. *British Weather and the Climate of Enlightenment.* Chicago: University of Chicago Press, 2007.

Golinski, Jan. *Making Natural Knowledge: Constructivism and the History of Science.* 2nd ed. Chicago: University of Chicago Press, 2005.

Gomperz, Theodor. *Die Apologie der Heilkunst: Hippocratis de Arte.* Leipzig: Veit & Comp., 1910.

Gottfried, Robert S. *The Black Death.* London: Robert Hale, 1983.

Grant, Hardy. "Leibniz and the Spell of the Continuous." *College Mathematics Journal* 25, no. 4 (September 1994): 291–294.

Gullan-Whur, Margaret. *Within Reason: A Life of Spinoza.* London: Jonathan Cape, 1998.

Hacking, Ian. *The Emergence of Probability: A Philosophical Study of Early Ideas About Probability, Induction, and Statistical Inference.* New York: Cambridge University Press, 2006.

Hacking, Ian. *Representing and Intervening.* Cambridge, UK: Cambridge University Press, 1983.

Hadot, Pierre. *The Veil of Isis: An Essay on the History of the Idea of Nature.* Translated by Michael Chase. Cambridge, MA: Harvard University Press, 2006.

Hansen, James, et al., "Target Atmospheric CO_2: Where Should Humanity Aim?" *Open Atmospheric Science Journal* 2 (2008): 217–231.

174   *Bibliography*

Harrison, Paul. "Theological Notes: A Promising Time for Pantheism." *The Independent*, May 21, 1999.

Harrison, Peter. "Was Newton a Voluntarist?" In *Newton and Newtonianism: New Studies*, edited by James E. Force and Sarah Hudson. Dordrecht: Kluwer, 2004, pp. 39–64.

Harrison, Peter. "Was There a Scientific Revolution?" European Review 15 (2007): 446–457.

Hartman, Chester and Gregory D. Squires. *There Is No Such Thing as a Natural Disaster: Race, Class, and Hurricane Katrina*. New York: Routledge, 2006.

Hartwig, G[eorg]. *Volcanoes and Earthquakes: A Popular Description of the Movements in the Earth's Crust: From "The Subterranean World."* London: Longmans Green, 1887.

Hayles, N. Katherine, ed. *Chaos and Order: Complex Dynamics in Literature and Science*. Chicago: University of Chicago Press, 1991.

Hayles, N. Katherine. *Chaos Bound: Orderly Disorder in Contemporary Literature and Science*. Ithaca, NY: Cornell University Press, 1990.

Hayles, N. Katherine. *How We Became Posthuman: Virtual Bodies in Cybernetics, Literature, and Informatics*. Chicago: University of Chicago Press, 1999.

Hegel, G.W.F. *Lectures on the History of Philosophy 1825–6*. Vol. III, *Medieval and Modern Philosophy*. Rev. ed. Translated and edited by George F. Brown. New York: Oxford University Press, 2009.

Heyd, Thomas, ed. *Recognizing the Autonomy of Nature: Theory and Practice*. New York: Columbia University Press, 2005.

Hippocrates. *The Art* [*On the Techne*], in *Hippocrates*. Edited and translated by W.H.S. Jones. London: William Heinemann, Loeb Classical Library, 1959, vol. 2, pp. 185–217.

Hobbes, Thomas. *Leviathan*. Aalen, Germany: Scientia, 1966 [1651].

Høffding, Harald. *A History of Modern Philosophy*. Translated by B.E. Meyer. London: MacMillan, 1924.

Holbach, Paul-Henry Thiry, Baron d.' *The System of Nature* (1770). Translated by H.D. Robinson. Boston: J.P. Mendum, 1853.

Hooke, Robert. *Lectures and Discourses of Earthquakes and Subterraneos Eruptions* (1668). In *The Posthumous Works of Robert Hooke*, containing his Cutlerian Lectures and other Discourses read at the Meetings of the Illustrious Royal Society. Published by Richard Waller. London: Printed by Sam Smith and Benj. Walford, (printers to the Royal Society) at the Princes Arms in St. Paul's Church-yard, 1705.

Hopkins, Jasper. *A Concise Introduction to the Philosophy of Nicholas of Cusa*. Minneapolis, MN: A.J. Banning Press, 1986.

Hubbeling, [H]ubertus.G. "Spinoza's Life: A Synopsis of the Sources and Some Documents." *Giornale critico della filosofia italiana* 8, nos. 3–4 (1977): 390–409.

Hume, David. *Essays: Moral, Political, and Literary*. Oxford, UK: Oxford University Press, 1963.

Hunt, John. *An Essay on Pantheism*. London: Longmans, Green, Reader, and Dyer, 1866.

Huxley, Thomas Henry. "On the Advisableness of Improving Natural Knowledge." A Lay Sermon delivered in St. Martin's Hall on Sunday, January 7th, 1866, and subsequently published in the "Fortnightly Review," p. 11 (1866). In *Collected Essays* (New York: D. Appleton, 1896), vol. 1.

Israel, Jonathan. *Enlightenment Contested: Philosophy, Modernity, and the Emancipation of Man, 1670–1752*. Oxford, UK: Oxford University Press, 2006.

Israel, Jonathan. *Radical Enlightenment: Philosophy and the Making of Modernity, 1650–1750*. Oxford, UK: Oxford University Press, 2001.

Iverach, James. *Descartes, Spinoza, and the New Philosophy*. New York: Scribner's, 1904.

Jacob, James R. *The Scientific Revolution: Aspirations and Achievements*. Atlantic Highlands, NJ: Humanities Press, 1998.

Bibliography 175

Jacob, Margaret C. *The Radical Enlightenment: Pantheists, Freemasons, and Republicans.* London: Allen & Unwin, 1981.

Jantsch, Erich. *The Self-Organizing Universe.* New York: Pergamon Press, 1980.

"John Scotus Erigena." http://www.britannica.com/EBchecked/topic/191466/John-Scotus-Erigena.

Jonson, Ben. *The Alchemist.* Edited with an Introduction and notes by Charles Montgomery Hathaway, Jr. Yale Studies in English, Albert S. Cook, editor. New York: Henry Holt, 1903.

Jonson, Ben. "Mercury Vindicated from the Alchemists at Court." *Works.* 11 vols. Oxford, UK: Clarendon, 1954–1965, vol. 7, pp. 407–417.

Kaebnick, Gregory, ed. *The Ideal of Nature: Debates About Biotechnology and the Environment.* Baltimore, MD: Johns Hopkins University Press, 2011.

Kautz, Richard. *Chaos: The Science of Predictable Random Motion.* New York: Oxford University Press, 2011.

Kayser, Rudolf. *Spinoza: Portrait of a Spiritual Hero.* New York: Philosophical Library, 1946.

Kellert, Stephen. *In the Wake of Chaos: Unpredictable Order in Dynamical Systems.* Chicago: University of Chicago Press, 1992.

Kendrick, T[homas] D[owning]. *The Lisbon Earthquake.* London: Methuen, 1956.

Klein, Naomi. *This Changes Everything: Capitalism vs. The Climate.* New York: Simon and Schuster, 2014.

Krüger, Lorenz, Lorraine Daston, and Michael Heidelburger, eds. *The Probablistic Revolution.* 2 vols. Cambridge, MA: MIT Press, 1987.

Kubrin, David. "How Sir Isaac Newton Helped Restore Law 'n' Order to the West." *Liberation* 16, no. 10 (Mar. 1972): 32–41.

Kuhn, Thomas. *The Structure of Scientific Revolutions.* Chicago: University of Chicago Press, 1962.

Latour, Bruno. *We Have Never Been Modern.* Cambridge, MA: Harvard University Press, 1993.

Leask, Ian. "Unholy Force: Toland's Leibnizian 'Consummation' of Spinozism." *British Journal for the History of Philosophy* 20, no. 3 (2012): 499–537.

Lehoux, Daryn. "Laws of Nature and Natural Laws." *Studies in History and Philosophy of Science* 37 (2006): 527–549.

Leibniz, Gottfried Wilhelm. *Der Briefwechsel von Gottfried Wilhelm Leibniz mit Mathematikern.* Edited by C. I. Gerhardt. Berlin: Mayer & Müller, 1899.

Leibniz, Gottfried Wilhelm. *Philosophical Papers and Letters.* Edited by Leroy Loemker. 2 vols. Chicago: University of Chicago Press, 1956.

Leibniz, Gottfried Wilhelm. *Theodicy.* Translated by E. M. Huggard. Gutenberg eBook, 2005.

Lermond, Lucia. *The Form of Man: Human Essence in Spinoza's Ethic.* Leiden: E. J. Brill, 1988.

Linsky, Amelia Kikue. *The Ferrara Earthquakes, 1570–1579: Science, Religion, and Politics in Late Renaissance Italy.* Bachelor of Arts in History. Middlebury, VT: Middlebury College, April 22, 2013.

Lloyd, Genevieve. *Part of Nature: Self-Knowledge in Spinoza's Ethics.* Ithaca, NY: Cornell University Press, 1994.

Lloyd, Genevieve. "Spinoza's Environmental Ethics." *Inquiry* 23, no. 3 (1980): 293–311.

Locke, John. *Two Treatises of Government.* Edited by Thomas I. Cook. New York: Hafner Press, 1973 [1690].

Lorenz, Edward. *The Essence of Chaos.* Seattle: University of Washington Press, 1993.

Lorenz, Edward. "Predictability: Does the Flap of a Butterfly's Wings in Brazil Set off a Tornado in Texas?" (1972), reprinted in Lorenz, Edward. *The Essence of Chaos.* Seattle: University of Washington Press, 1993.

176   Bibliography

Lovejoy, Arthur O. *The Great Chain of Being: A Study in the History of an Idea.* Cambridge, MA; Harvard University Press, 1936.

Lovejoy, Arthur O. and George Boas. *Primitivism and Related Ideas in Antiquity.* New York: Octagon Books, 1965.

Lucas, Jean Maximilien. "The Life of the Late M. de Spinoza." Edited and translated by A[braham] Wolf. *The Oldest Biography of Spinoza.* London: Allen & Unwin, 1927.

Lucas, [Jean Maximilien]. *La Vie de M. Benoit de Spinoza* (1735). In *Vies de Spinoza.* Paris: Allia, 1999.

Lucks, Henry A. "*Natura Naturans—Natura Naturata.*" *The New Scholasticism* 9, no. 1 (1935): 1–24.

Lucretius. *De Rerum Natura* (1st c. B.C.E.). Translated by W. E. Leonard. New York: E. P. Dutton, 1950.

Lucretius. *De Rerum Natura* (Latin). http://www.thelatinlibrary.com/lucretius.html

Lucretius. *The Nature of the Universe* (*De Rerum Natura*, 1st c. B.C.E.). Translated by R. E. Latham. Baltimore, MD: Penguin, 1951.

Lucretius. *On the Nature of the Universe* (*De Rerum Natura*). A New Verse Translation, with an Introduction by James H. Mantinband. New York: Frederick Ungar, 1965.

Luig, Klaus. "Leibniz's Concept of *Jus Naturale* and *Lex Naturalis*—Defined 'with Geometric Certainty.'" In *Natural Law and Laws of Nature in Early Modern Europe*, edited by Lorraine Daston and M. Stolleis. Burlington, VT: Ashgate, 2008, pp. 183–198.

MacDonald, Michael, ed. *Witchcraft and Hysteria in Elizabethan London: Edward Jorden and the Mary Glover Case.* London: Tavistock/Routledge, 1991.

Marris, Emma. *Rambunctious Garden: Saving Nature in a Post-Wild World.* New York: Bloomsbury, 2011.

Martin, Craig. *Renaissance Meterology: Pomponassi to Descartes.* Baltimore, MD: Johns Hopkins University Press, 2011.

Mathews, Freya. *The Ecological Self.* London: Routledge, 1991.

Mauch, Christof and Christian Pfister, eds. *Natural Disasters, Cultural Responses: Case Studies Toward a Global Environmental History.* Lanham, MD: Lexington Books, 2009.

McKibben, Bill. "The Arctic Ice Crisis: Greenland's Glaciers Are Melting Far Faster than Scientists Expected." *Rolling Stone*, August 30, 2012.

McKibben, Bill. *The End of Nature.* New York: Random House, 1989.

McKibben, Bill. "Global Warming's Terrifying New Math." *Rolling Stone*, August 2, 2012.

McKibben, Bill. "The Keystone Pipeline Revolt: Why Mass Arrests Are Just the Beginning." *Rolling Stone*, October 13, 2011.

McPeek, John. *Bruno, Spinoza, and the Terms* Natura Naturans *and* Natura Naturata. New Orleans: Tulane University, 1965.

McPhee, John. *The Control of Nature.* New York: Farrar, Straus, and Giroux, 1989.

Megenberg, Konrad von. *Buch der Natur.* Edited by Franz Pfeiffer. Stuttgart: Karl Aue, 1861.

Melamed, Yitzhak. "Spinoza's Metaphysics of Substance: The Substance-Mode Relation as a Relation of Inherence and Predication." *Philosophy and Phenomenological Research* 78, no. 1 (January 2009): 17–82.

Mendelson, Michael. "Saint Augustine." *The Stanford Encyclopedia of Philosophy.* http://plato.stanford.edu/entries/augustine/. 2010.

Merchant, Carolyn. "The BP Oil Spill: Economy Versus Ecology." American Society for Environmental History. *ASEH News* 21, no. 2 (Summer 2010), http://aseh.net/publications/newsletter.

Merchant, Carolyn. *The Death of Nature: Women, Ecology, and the Scientific Revolution.* San Francisco: HarperCollins, 1980; 2nd ed. 1990.

Merchant, Carolyn. *Earthcare: Women and the Environment.* New York: Routledge, 1996.

Bibliography   177

Merchant, Carolyn. "Francis Bacon and the 'Vexations of Art': Experimentation as Intervention." *British Journal for the History of Science* 46, no. 4 (2013): 551–599.

Merchant, Carolyn, ed. *Key Concepts in Critical Theory: Ecology*. 2nd ed. Amherst, NY: Humanity Books, 2008.

Merchant, Carolyn. *Radical Ecology: The Search for a Livable World*. New York: Routledge, 2006.

Merchant, Carolyn. *Reinventing Eden: The Fate of Nature in Western Culture*. New York: Routledge, 2003; 2nd ed. 2013.

Merton, Thomas. "Hagia Sophia," *Dawn. The Hour of Lauds. Collected Poems*. New York: New Directions, 1977.

Miller, Clyde Lee. "Cusanus, Nicolaus [Nicolas of Cusa]." The Stanford Encyclopedia of Philosophy (Fall 2009 edition). Edward N. Zalta, ed. http://plato.stanford.edu/archives/fall2009/entries/cusanus/.

Miller, Jon. "Spinoza and the Concept of a Law of Nature." *History of Philosophy Quarterly* 20, no. 3 (July 2003): 257–276.

Mitchell, Sandra. *Unsimple Truths: Science, Complexity, and Policy*. Chicago: University of Chicago Press, 2009.

Mooney, Hilary Anne-Marie. *Theophany: The Appearing of God According to the Writings of Johannes Scotus Eriugena*. Tubingen, Germany: Mohr Siebeck, 2009.

Moran, Dermot. "Pantheism in John Scottus Eriugena and Nicholas of Cusa." *American Catholic Philosophical Quarterly* (Winter 1990): 131–152.

Moran, Dermot. *The Philosophy of John Scottus Eriugena. A Study of Idealism in the Middle Ages*. Cambridge, UK: Cambridge University Press, 1989.

Naddaf, Gerard. *The Greek Concept of Nature*. Albany: State University of New York Press, 2005.

Nadler, Steven. *Spinoza: A Life*. Cambridge, UK: Cambridge University Press, 1999.

Naess, Arne. *Ecology of Wisdom: Writings by Arne Naess*. Edited by Alan Drengson and Bill Devall. Berkeley, CA: Counterpoint Press, 2008.

Naess, Arne. "Environmental Ethics and Spinoza's *Ethics*: Comments on Genevieve Lloyd's Article." *Inquiry* 23, no. 3 (1980): 313–325.

Naess, Arne. *The Selected Works of Arne Naess*. Edited by Harold Glasser and Alan Drengson. 10 vols. Dordrecht, Netherlands: Springer, 2005.

Naess, Arne. "The Shallow and the Deep, Long-Range Ecology Movement: A Summary." *Inquiry* 16, no. 1 (1973): 95–100.

Naess, Arne. "Spinoza and Ecology." *Philosophia* 7 (1977): 45–54.

Naess, Arne. *Spinoza and the Deep Ecology Movement*. Delft, Netherlands: Eburon, 1993.

Newman, William R. *Atoms and Alchemy: Chymistry and the Experimental Origins of the Scientific Revolution*. Chicago: University of Chicago Press, 2006.

Newman, William R. *Promethean Ambitions: Alchemy and the Quest to Perfect Nature*. Chicago: University of Chicago Press, 2004.

Newman, William R. and Anthony Grafton, eds. *Secrets of Nature: Astrology and Alchemy in Early Modern Europe*. Cambridge MA: MIT Press, 2001.

Newman, William R. and Lawrence M. Principe. *Alchemy Tried in the Fire: Boyle and the Fate of Helmontian Chymistry*. Chicago: University of Chicago Press, 2002.

Newton, Isaac. *Mathematical Principles of Natural Philosophy*. Translated by Andrew Motte (1729), rev. Florian Cajori. Berkeley: University of California Press, 1960.

Newton, Isaac. *Opticks*. Based on 4th ed. 1730. New York: Dover, 1952.

Niemi, Jari. "Spinoza's Political Philosophy." *Internet Encyclopedia of Philosophy*. http://www.iep.utm.edu/spin-pol/.

Noone, Tim and R.E. Hauser. "Saint Bonaventure." (2013). http://plato.stanford.edu/entries/bonaventure/.

178   Bibliography

O'Meara, John J. and Ludwig Bieler, eds. *The Mind of Eriugena.* Dublin: Irish University Press, 1973.

Otten, Willemien. *The Anthropology of Johannes Scottus Eriugena.* Leiden: E.J. Brill, 1991.

Otten, Willemien. "The Dialectic of the Return in Eriugena's *Periphyseon.*" *Harvard Theological Review* 84, no. 4 (1991): 399–421.

Paice, Edward. *Wrath of God: The Great Lisbon Earthquake of 1755.* London: Quercus, 2008.

Panofsky, Erwin. *Renaissance and Renascences in Western Art.* Stockholm: Almqvist and Wiksell, 1960.

Paracelsus. *The Book Concerning the Tincture of the Philosophers* . . . Transcribed by Dusan Djordjevic Mileusnic from *Paracelsus his Archidoxis: Comprised in Ten Books, Disclosing the Genuine way of making Quintessences, Arcanums, Magisteries, Elixirs, &c. Together with his Books Of Renovation & Restauration.* London: J.H. Oxon, 1660.

Paracelsus. *The Coelum Philosophorum, or Book of Vexations,* in *The Hermetic and Alchemical Writings of Aureolus Philippus Theophrastus Bombast, of Hohenheim, called Paracelsus the Great* . . . . Edited and translated by Arthur Edward Waite. 2 vols. Berkeley, CA: Shambala Books, 1976.

Park, Katharine. *Secrets of Women: Gender, Generation, and the Origins of Human Dissection.* Cambridge, MA: Zone Books, 2006.

Park, Katharine and Lorraine Daston. *Wonders and the Order of Nature, 1150–1750.* Cambridge, MA: Zone Books, 1998.

Parker, Geoffrey. *Global Crisis: War, Climate Change and Catastrophe in the Seventeenth Century.* New Haven, CT: Yale University Press, 2013.

Pelling, Mark and Juha I. Uitto. "Small Island Developing States: Natural Disaster Vulnerability and Global Change." *Environmental Hazards* 3 (2001): 49–62.

Penna, Anthony and Jennifer S. Rivers. *Natural Disasters in a Global Environment.* New York: Wiley Blackwell, 2013.

Picard, Rosalind W. "Newton: Rationalizing Christianity or Not." http://web.media.mit.edu/~picard/personal/Newton.php.

Plato. *The Dialogues of Plato.* Translated by Benjamin Jowett. New York: Random House, 1937 [1892]). "Timaeus," "Critias."

Pliny the Younger. *The Letters of the Younger Pliny.* Translated with an Introduction by Betty Radice. Baltimore: Penguin Books, 1969.

Plotinus. *Enneads.* Translated by Stephen Mackenna. 5 vols. London: Medici Society, 1917.

Plotinus. *Enneads.* In *Collected Writings of Plotinus.* Translated by Thomas Taylor. Frome, Somerset, UK: Prometheus Trust, 1994.

Plumptre, Constance E. *General Sketch of the History of Pantheism.* 2 vols. London: Beacon, 1878–1879.

Plumwood, Val. *Feminism and the Mastery of Nature.* London: Routledge, 1993.

Pollock, Frederick. *Spinoza.* London: Duckworth, 1935.

Popkin, Richard H. "Hume and Spinoza." *Hume Studies* 5, no. 2 (November 1979): 65–93.

Prigogine, Ilya and Isabelle Stengers. *Order out of Chaos: Man's New Dialogue with Nature.* Introduction by Alvin Toffler. New York: Bantam, 1984.

Principe, Lawrence M. *The Secrets of Alchemy.* Chicago: University of Chicago Press, 2013.

Radkau, Joachim. *Nature and Power: A Global History of the Environment.* Translated by Thomas Dunlap. New York: Cambridge University Press, 2008.

Rees, Graham. "Francis Bacon's Semi-Paracelsian Cosmology." *Ambix* 12, Pt. 2 (July 1975): 81–101.

Robin, Libby, Sverker Sörlin, and Paul Warde, eds. *The Future of Nature: Documents of Global Change.* New Haven, CT: Yale University Press, 2013.

Bibliography 179

Rochberg, Francesca. *The Heavenly Writing: Divination, Horoscopy and Astronomy in Mesopotamian Culture*. New York: Cambridge, 2004.

Rochberg, Francesca. *In the Path of the Moon: Babylonian Celestial Divination and Its Legacy*. Leiden: Brill, 2010.

Rohr, Christian. "Man and Natural Disaster in the Late Middle Ages: The Earthquake in Carinthia and Northern Italy on 25 January 1348 and Its Perception." *Environment and History* 9 (2003): 127–149.

Rossi, Charlotte. "How Are Natural Disasters Socially Constructed?" http://www.studymode.com/essays/How-Are-Natural-Disasters-Socially-Constructed-684217.html.

Rutherford, Donald. "Spinoza's Conception of Law." In *Spinoza's Theological–Political Treatise: A Critical Guide*, edited by Yitzhak Y. Melamed and Michael A. Rosenthal. New York: Cambridge University Press, 2010, Ch. 8, pp. 143–167.

Rutkow, Ira M. *Surgery: An Illustrated History*. London and Southampton: Elsevier Science Health Science, 1993.

"Saint Thomas Aquinas." http://www.themiddleages.net/people/aquinas.html.

Scruton, Roger. *Spinoza: A Very Short Introduction*. Oxford, UK: Oxford University Press, 2002 [1986].

Sessions, George. "Spinoza and Jeffers on Man and Nature." *Inquiry* 20 (1977): 481–528.

Shapin, Steven and Simon Schaffer. *Leviathan and the Air Pump: Hobbes, Boyle, and the Experimental Life*. Princeton, NJ: Princeton University Press, 1985.

Siebeck, H[ermann]. "Über die Enstehung der Termini *natura naturans* und *natura naturata*." *Archiv für Geschichte der Philosophie* 3 (1890): 370–378.

Silvestris, Bernard. *Cosmographia*. Edited by Peter Dronke. Leiden: E. J. Brill, 1978.

Simberloff, Daniel. "A Succession of Paradigms in Ecology: Essentialism to Materialism and Probabilism." *Synthese* 43, no. 1 (January 1980): 3–39.

Singer, Dorothea Waley. *Giordano Bruno: His Life and Thought, with Annotated Translation of His Work*, On the Infinite Universe and Worlds. New York: Schuman, 1950.

Snyder, Gary. *No Nature: New and Selected Poems*. New York: Pantheon, 1992.

Snyder, Laura J. *Reforming Philosophy: A Victorian Debate on Science and Society*. Chicago: University of Chicago Press, 2006.

Soper, Kate. *What Is Nature? Culture, Politics, and the Non-Human*. Oxford, UK: Blackwell, 1995.

Sörlin, Sverker, ed. *Nature's End: History and the Environment*. Palgrave MacMillan, 2009.

Spinoza, Baruch. *Ethics* (1677). Edited and translated by Edwin Curley. New York: Penguin, 1996.

Spinoza, Baruch. *Principles of Cartesian Philosophy*. Translated by Harry E. We. New York: Philosophical Library, 1961.

Spinoza, Baruch. *The Principles of Cartesian Philosophy and Metaphysical Thoughts*; Followed by Inaugural dissertation on matter; Lodewijk Meyer; translated by Samuel Shirley, Introduction and notes by Steven Barbone and Lee Rice. Indianapolis, IN: Hackett, 1998.

Spinoza, Baruch. *Principles of Descartes's Philosophy*. Translated by Halbert Hains Britan. LaSalle, IL: Open Court, 1961 [1663]. Reprint, 2010.

Spinoza, Baruch. *Selections*. Edited by John Wild. New York: Scribner's, 1930.

Spinoza, Baruch. *Spinoza's Short Treatise on God, Man, and His Well-Being*. Translated and edited by A[braham] Wolf. London: Adam and Charles Black, 1910.

Spinoza, Baruch. *Theological-Political Treatise*. Edited by Jonathan Israel. Translated by Michael Silverthorne and Jonathan Israel. New York: Cambridge University Press, 2007.

Spinoza, Benedict de. *A Theologico-Political Treatise and A Political Treatise*. Translated by R.H.M. Elwes. London: George Bell, 1900 [1883]. Reprint, New York: Dover, 1951; Cosimo, 2007.

180 Bibliography

Spinoza, Benedictus de. *Spinoza Opera*. Edited by Karl Gebhardt. 4 vols. Heidelberg: Carl Winter, 1925.

Stewart, Matthew. *The Courtier and the Heretic: Leibniz, Spinoza and the Fate of God in the Modern World*. New York: W.W. Norton, 2006.

Stigler, Stephen M. *The History of Statistics: The Measurement of Uncertainty Before 1900*. Cambridge, MA: Harvard University Press, 1986.

Stimson, Dorothy. *Scientists and Amateurs: A History of the Royal Society*. New York: Schuman, 1948.

Swan, John. *A True and Brief Report, of Mary Glovers Vexation, and of her Deliverance by the Means of Fasting and Prayer*. In *Witchcraft and Hysteria in Elizabethan London: Edward Jorden and the Mary Glover Case*. Edited with an Introduction by Michael MacDonald. London: Tavistock/Routledge, 1991.

"Thomas Aquinas." http://en.wikipedia.org/wiki/Thomas_Aquinas.

Thompson, Charles John Samuel. *Alchemy and Alchemists*. Mineola, NY: Dover, 2002 [1932].

Thomson, Ann. *Bodies of Thought: Science, Religion, and the Soul in the Early Enlightenment*. New York: Oxford University Press, 2008.

Thrower, Norman J.W., ed. *Standing on the Shoulders of Giants: A Longer View of Newton and Halley*. Berkeley: University of California Press, 1990.

Tiles, Mary. "Experiment as Intervention." *British Journal for the Philosophy of Science* 44 (1993): 463–475.

Tillyard, Eustace M.W. *The Elizabethan World Picture*. New York: Vintage, 1959.

Toland, John. *Christianity not Mysterious or, A treatise shewing that there is nothing in the Gospel contrary to reason, nor above it: and that no Christian doctrine can be properly call'd a mystery*. London: Printed for Sam Buckley, 1696.

Toland, John. *A Collection of Several Pieces of Mr. John Toland, Now first published from his Original Manuscripts, With Some Memoirs of his Life and Writings*. 2 vols. London: Printed for J. Peale, 1726.

Toland, John. *Letters to Serena: containing, I. The origin and force of prejudices. II. The history of the soul's immortality among the heathens. III. The origin of idolatry, and reasons of heathenism, As also, IV. A letter to a gentleman in Holland, showing Spinosa's system of philosophy to be without any principle or foundation. V. Motion essential to matter; an answer to some remarks by a nobel friend on the confutation of Spinosa. To all which is prefixed, VI. a preface; being a Letter to a gentleman in London*. London: Printed for M. Cooper, W. Reeve, and C. Sympson, 1753.

Toland, John. *Pantheisticon, sive formula celebrandae sodalitatis socraticae* (London, 1720); *Pantheisticon, or, the form of celebrating the Socratic-Society. Divided into three parts. . . . To which is prefix'd a discourse upon the antient and modern societies of the learned, . . . And subjoined, a short dissertation upon a two-fold philosophy of the pantheists, . . . Written originally in Latin, by the ingenious Mr. John Toland. And now, for the first time, faithfully rendered into English*. London: Printed for Sam. Paterson; and sold by M. Cooper, 1751.

Torchia, N[atale] Joseph. *Plotinus, Tolma, and the Descent of Being*. New York: Peter Lang, 1993.

Tuchman, Barbara W. *A Distant Mirror: The Calamitous Fourteenth Century*. New York: Knopf, 1978.

Vallee, Gerard, trans. *The Spinoza Conversations Between Lessing and Jacobi, Text with Excerpts from the Ensuing Controversy*. Introduction by Gerard Vallee. Translated by G. Vallee, J.B. Lawson, and C.G. Chapple. Lanham, MD: University Press of America, 1988.

Van Liere, Frans. "The Latin Bible: c. 900 to the Council of Trent, 1546." In *From 600 to 1450. The New Cambridge History of the Bible*, vol. 2, edited by Richard Marsden and E. Ann Matter. Cambridge, UK: Cambridge University Press, 2012, pp. 93–109.

Virgil. *Aeneid*. Translated by Patricia Johnson. Norman: University of Oklahoma Press, 2012.

Virgil. *The Eclogues*. The Latin text with a verse translation and notes by Guy Lee. Rev ed. New York: Penguin, 1984.

Virgil. *Georgics: A Poem of the Land*. Translated and edited by Kimberly Johnston. New York: Penguin, 2009.

Virgil. Latin texts. http://www.thelatinlibrary.com/verg.html.

Von Staden, Heinrich. "The Discovery of the Body: Human Dissection and Its Cultural Contexts in Ancient Greece." *Yale Journal of Biology and Medicine* 65 (1992): 223–241.

Von Staden, Heinrich. "*Physis* and *Techne* in Greek Medicine." In *The Artificial and the Natural: An Evolving Polarity*, edited by Bernadette Bensaude-Vincent and William R. Newman. Cambridge, MA: MIT Press, 2007, pp. 21–49.

Waldrop, M. Mitchell. *Complexity: The Emerging Science at the Edge of Order and Chaos*. New York: Simon and Schuster, 1992.

Wallis, John. "A Defence of the Royal Society, and the Philosophical Transactions, particularly those of July 1670 . . . in a Letter to the Right Honourable William Lord Viscount Brouncker, March 6, 1678. London: Thomas Moore, 1678.

Weber, Alfred. *History of Philosophy*. Translated by Frank Thilly from the 5th French edition. New York: Charles Scribner's Sons, 1896.

Weijers, Olga. "Contribution à l'histoire des termes '*nautura naturans*' et '*natura naturata*' jusqu'à Spinoza." *Vivarium* 16, no. 1 (1978): 70–80.

Weinberg, Joanna. "'The Voice of God': Jewish and Christian Responses to the Ferrara Earthquake of November 1570." *Italian Studies* 46 (1991): 69–81.

Weisman, Alan. *The World Without Us*. New York: Picador Press, 2007.

Weisner, Merry, Julius R. Ruff, and William Bruce Wheeler. *Discovering the Western Past*. Boston: Houghton Mifflin, 1989.

Wells, Jennifer. *Complexity and Sustainability*. New York: Routledge, 2013.

Wells, Jennifer and Carolyn Merchant. "Melting Ice: Climate Change and the Humanities." *Confluence* XIV, no. 2 (Spring 2009): 13–27.

West, John B. "Robert Boyle's Landmark Book of 1660 with the First Experiments on Rarified Air." *Journal of Applied Physiology* 98, no. 1 (January 2005): 31–39.

Westfall, Richard S. *The Construction of Modern Science: Mechanisms and Mechanics* New York: Wiley, 1971.

Westfall, Richard S. *Force in Newton's Physics. The Science of Dynamics in the Seventeenth Century*. New York: Elsevier, 1971.

Westfall, Richard S. *Never at Rest: A Biography of Isaac Newton*. New York: Cambridge University Press, 1980.

Westman, Robert and David Lindberg, eds. *Reappraisals of the Scientific Revolution*. Cambrige, UK: Cambridge University Press, 1990.

White, Jr. Lynn. Medieval Technology and Social Change. Oxford: Clarendon Press, 1962.

Whitney, Charles. *Francis Bacon and Modernity*. New Haven, CT: Yale University Press, 1986.

Williams, Raymond. *Keywords*. 2nd ed. New York: Oxford University Press, 1985 [1976], "Nature," pp. 219–224.

Woolhouse, R[oger] S. *Starting with Leibniz*. London: Continuum, 2010.

Worsham, Lynn and Gary A. Olson, eds. *Plugged In: Technology, Rhetoric, and Culture in a Posthuman Age*. Cresskill, NJ: Hampton Press, 2008.

Index

Page numbers in bold indicate tables.

Accademia del Cimento, Italy 91
active principle of God (Aquinas), 52–3, 56
active principles (Newton): God is present
through, 135–37, 139–40; gravity as
expressing, 136, 139, 141; manifest
in fermentation, 135–36, 139, 141; as
reinvigorating the world, 126
activity of God (Augustine), 45; (Spinoza),
107, 109
activity of nature, 29, 31, 32, 34, 44, 154
Adam, Barbara, *Timescapes of Modernity*
133–34
Advancement of Learning, The (Bacon) 84,
85–6
air pump (Guerike) 91–2
Alaska, climate change effects in 149, 164
Alberti, Leon Battista, **12**, 70
Alchemist, The (Jonson) 88
alchemy 86–8
Ames, Christine Caldwell, *Righteous
Persecution* 88
Analytic Theory of Probabilities (LaPlace)
9, 151
animal experimentation 92–3, *93*
anthrax 65
Anthropocene 6, 154–55
anthropocentric ethical systems 161–62
Aquinas, Thomas, **12**, 30, 44, 50; on
active principle of God, 52–3, 56;
commentator on Aristotle's works,
50; on the corruptibility of matter,
51, 52; *De Principiis Naturae* ("On the
Principles of Nature"), 51; on God's
eternal law, 52, 53; on generation,
51, 52; on material, efficient, formal
and final causes, 51–2; on natural

disasters, 52; on natural evils, 52;
on natural law, 52–3, 96; on *natura
naturans* as active power, 50–51, 53,
54, 56; rationality in, 44, 45, 49, 51–3,
56; *Summa Theologica*, 51–2, 109;
Thomism and, 110
Aristotle, 22, 23; chance vs. necessity in,
30; influence of on Aquinas, 50; on
material, formal, efficient, and final
causes, 29–30; *Physis* (*Physics*), 24, 25,
29, 39, 43, 50, 51, 67; on *physis* and
techne, 29
Ars Conjectandi (Bernoulli) 151
artists and artisans 70–71
atomism: Greco-Roman, **12**, 30; particle
theory and, 113, 132
Augmentis, De (Bacon) 28–9
Augustine, **12**; on humans and animals,
45–6; influence of on Bonaventure,
54; merging Christianity with
Neoplatonism, 44–5
autonomous nature. *See* nature as
autonomous; partnership ethic
Averroes, *Commentary* of, 50

Bacon, Francis, **12**, 56, *83*; on the binding
of and inquisition of nature, 89–90,
91; *De Augmentis*, 28–9; "History of
Winds", 94; *Instauratio Magna*, 28;
myth of Proteus, 28, 82, 87, 88; *natura
naturans/natura naturata* relation
in, 83, 85–6, 90, 95–6; on "nature
naturing"/"nature natured", 84–6;
Novum Organum, 28, 84, 85; on *techne*
(art), 83, 84, 86; *The Advancement of
Learning*, 84, 85–6; *Thema Coeli*, 87;

184   *Index*

on the three states of nature, 85–6, 87; on vexation, 82, 83–4, 86–90 (*see also* experimentation; vexing); *Wisdom of the Ancients*, 28–9

Being, the nature of 24, 31, 34

bell jar experiments 92–3, *93*

Bentham, Jeremy 162

Bentley, Richard 139

Bernoulli, Jacob, 4, 9; *Ars Conjectandi*, 151

Bialostocki, Jan, 70

Bible, 117; Book of Genesis, 47, 54; King James version of, 88; Isaiah 88; Leviticus 67; Psalms 67

biodiversity 5, 155, 162

biological systems 140, 152–53. *See also* complex systems

biotechnology, **12**

Black Death outbreaks, 12t, 64, *64;* Bubonic Plague of 1348, 11, **12**, 63–4, *64*, 132; Great Plague of London of 1665, 132, *133*

Blumenberg, Hans 69–70

Boccaccio, Giovanni, *Decameron* 63

Boltzmann, Ludwig, 9, 151

Bonaventure, **12**, 37, 44, 53–4; *Commentaries on . . . Peter Lombard*, 54; on the corruption of humanity through the will, 56; on the hierarchies of the creation, 55; on *natura naturans* and *natura naturata*, 50, 54, 61n.37–9; on the Trinity of God, 54–5

Book Concerning the Tincture of the Philosophers (Paracelsus) 87–8

Botkin, Daniel, 6, 37; *The Moon in the Nautilus Shell*, 4–5

Boyle, Robert, **12**, 76, 84, *91*, 113, 158; *New Experiments Physico-Mechanicall. . . . .* , 91–2; *Robert Boyle's Air Pump*, *92*

Bruno, Giordano: bridging medievalism and Enlightenment science, 73–5; *Cause, Principle and Unity*, 74, 109; cosmic nature in, 73–4, 109; imprisoned and burned, 75; *On the Monad; Number and Figure* (*De Monade, numero et figura*), 75

Bubonic Plague of 1348, 11, **12**, 63–4, *64*, 132. *See also* Black Death outbreaks

butterfly effect (Lorenz) 7–8, 152

Cabinet of Nature's Secrets 86, 87

calculus 9, 14, 97, 113, 140; calculating machine, 128; discovery of (Leibniz), 125, 128, 129–30, 131; Newtonian, 132

Calvus, Marcus Fabius 29, 38n17, 84

capitalism: and nature, 159–60; preindustrial, 66–7

carbon emissions: burning of fossil fuels, 5, 6, 149–50, 154–55; upper limit of, 155. *See also* climate change

Cartesian dualism, 105, 132, 157; Spinoza responding to, 103, 105–07, 120n.15

"Case for Complexity" (Mitchell) 5–6

Ceres asteroid 151

certainty 1, 7, 26, 32, 72, 103, 105–07, 134, 152–53; laws of nature as (Spinoza), 103; in mathematics (*see also* mathematics), 72; *Natura* allows for (Lucretius), 32–3; the problem of (*see also* laws of nature; predictability/ prediction), 105–06. *See also* uncertainty

chance vs. necessity (Aristotle) 30

change in nature, 103; Greco-Roman concepts of, 23, 26, 30; Heraclitus on, 24–5, 26; Leibniz on, 145n.14; Plotinus on, 34–7. *See also* chaos theory; complex systems

changelessness (permanence), 9, 25, 37. *See also* Demiurge; Monad

chaos: *Khaos* as the original creatrix (Hesiod), **12**, 13, 21–3, 26; the universe could have arisen from (Descartes), 106, 119n.113; as unpredictability (*see* unpredictability of nature). *See also* chaos theory, unpredictability; unruly and uncontrollable nature

chaos and complexity science, 10–11, **12**, 31, 140, 144, 150; as a new paradigm, 1, 2, 6, 14, 144, 151, 153, 157. *See also* chaos theory; complexity theory

Chaos: Making a New Science (Gleick) 151–52

chaos theory, 1, 7–8, 130, 144; the butterfly effect (Lorenz), 7–8, 152; emerging in the mid- Twentieth c., 10, 140; nonlinear descriptions of nature in, 151–52; rooted in a paradigm

of autonomous nature, 149, 150, 153, 161
Chaos: The Science of Predictable Random Motion (Kautz) 10–11
Châtelet, Gabrielle Emile du 131
Chester, David 42
Christianity, constructions of nature 42, 44
Christian Neoplatonism 44–5, 47, 69, 71, 73
Cicero, *De Legibus* 31, 96
City of God (Augustine) 44
civil society, 75, 95, 96, 104–05, 112. *See also* government
Clark, Robert David 32–3
Clarke, Samuel 137, 138; Leibniz-Clarke Correspondence, 137–40
class, social 25, 66, 177
climate change, 4, **12**, 14, 15n.14, 150; anticipated outcomes, 155–57; autonomous nature and, 149–50, 154–55; global temperature, 155; the partnership ethic and, 153, 164–65; randomness of events, 6–8
climate justice 164–65
clock metaphor of the world 5, 137–39, 140; clocklike universe 3, 10, 103 God as the clockmaker, 97;
closed systems: energy decreasing in (entropy), 140; mechanism and, 126–27, 131
code of ethical accountability 162
Coelum Philosophorum (Paracelsus) 86–7
Cogita Mathematica (Spinoza) 107
Cohen, Bernard 165–66n.5
Commentaries on . . . Peter Lombard (Bonaventure) 54
"common laws of the state" (Spinoza) 105
complexity theory xi, xii, xiii, 1, 5–6, 14, 144, 161; Daniel Botkin on, 5–6; Jennifer Wells on 3, 5, 152–53; as a new paradigm, 2, 151; prehistory of (Merchant) xi, 6, 14, 165; Sandra Mitchell on, 5–6
complex systems, 6–7, 10, 16, 160; biological systems, 140, 152–53; random aspect of, 161; as unstable systems, 152–53. *See also* uncertainty
computers and digital concepts 5, 12t, 159, 161

Concerning the Nature of Things (Paracelsus) 87
Confessions (Augustine) 44
conservation of energy, law of (Leibniz) 131
control of nature, 11, 13, 75–6, 79; binding nature through experimentation (Bacon), 81, 82–6, 87; as domination, 152–53; as a problem or question, 1, 7, 22, 31, 56, 150; through advances in science and mathematics (*see also* mechanism), 131, 144. *See also* nature as autonomous; unruly and uncontrollable nature
Copernicus, Nicholas 73, 96, 106
Cornarius, Janus 29, 38n.17, 84
corporeal bodies 131
corpuscles, material **12**, 30, 106
corruptibility of matter (Aquinas) 51–2
cosmos, descriptions of in different periods, **12**; sun-centered, 73–4, 96–7
Cosmographia (Silvestris) 68–9
creator: the Demiurge (Plato), 22, 25–7; distinguished from the created world (Augustine), 45; Natura Creatrix (Lucretius), **12**, 22, 31–3; nature as an active, 68, 70, 75; unity with creation (Erigena), 47–8; the unmoved mover (Aristotle), 30. *See also* God; *natura/ naturans*
Crete, volcanic disaster 1, **12**, 14n.1
crop failure and famine 155
cultural diversity 162
Cusa, Nicholas of, **12**, 69, 73, 109; *On Learned Ignorance (De Docta Ignorantia)*, 69–70
cybernetics 150, 160

d'Alembert, Jean 131
Darwinian evolution. *See* evolution
da Parma, Giovanni 42
da Vinci, Leonardo 67; "*Deluge*", 72; Garrard on, 70–72; "*The Great Lady*", 71; portrait of Cecilia Gallerani, 71; *Virgin of the Rocks*, 71
Davis, Edward 137, 146–47n.35
Decameron (Boccaccio) 63
decay, concern over universal (Newton), 135, 136–37, 139. *See also* entropy

186   *Index*

deep ecology 116–17
"*Deluge*" (Leonardo), 72
Demiurge (Plato), 22, 23, 25–7, 68, 134
Democritus, 30, 112
Demon, of Laplace and Maxwell 9–10,
    146n.21
Demonic Vexation *82. See also* Bacon;
    Satan; vexation
De Principiis Naturae (Aquinas) 51
De Rerum Natura (On the Nature of
    Things—Lucretius) 31
Descartes, René, 30, 84, 105; *cogito ergo
    sum* (I think therefore I am), 132; *Le
    Monde* (*The World*), 106; *Meditations
    on the First Philosophy*, 132; on
    physical vs. psychic beings, 105–6;
    Principles of Philosophy, 105, 135;
    that universe could have arisen from
    chaos, 106, 119n.13, 119–20n.14; on
    the vortex motion of corpuscles, **12**,
    106–7, 112. *See also* Cartesian dualism
desertification 155, 156. *See also* climate
    change
determinism,10, 101, 112; deterministic
    mechanics (*see also* mechanism, 9–10,
    140, 146n.21; deterministic systems,
    2, 7
Deus sive Natura (Spinoza) 103, 115–16
d'Holbach, Baron (Paul-Henri Thiry) 143
dialectics:) ; 66, 69 in Cusa; in Erigina, 24,
    32; 47
digital concepts and computers 5, 12t,
    159, 161
Dionysius the Areopagite 54
disease 4, 27, 28–9, 75
disease epidemics 21, 66–67, 108. *See also*
    Bubonic Plague; Black Death
disorder in nature *See* order and disorder;
    unruly and uncontrollable nature
Divine Craftsman (Plato) 22, 23, 25–7, 68,
    134. *See also* Demiurge
Divine Providence (*Noys*) 68
Dronke, Peter 68–9
drought 2, 93. *See also* desertification;
    natural disasters
dualistic conceptions; between (passive)
    matter and force, 134; between human
    and (non-human) nature, 159–60;
    creator separate from creation, 8,
    9, **12**, 13, 45, 70–71; passive matter

versus force (Newtonian), 134; pure
    forms vs. the world of appearances
    (Platonic), 9, 23, 25–6, 34, 160. *See
    also* Cartesian dualism
Dürer, Albrecht "Artist Drawing a Nude
    with Perspective Device" 72
dynamical systems. *See* complex systems

Earthcare (Merchant), 7–8
earth-centered cosmos (Ptolemy) 96, 151
earthly nature **12**
earthquakes, 2, 3, 11, 33, 42, 66, 93; Hooke's
    Law and, 141, 147n.40
earthquakes cited: Earthquake of 1348
    (Italy), **12**, *43*, 57n.1; earthquake of
    1570 (Italy), 67; early debate over
    causes of, 141; Lisbon Earthquake of
    1755, 11, **12**, 127–28, 140–43, *142*
ecocentric ethics 162
ecologically sound management 162
"Ecology, Nature, and Responsibility"
    (Soper) 159–60
efficient cause 29
Einstein, Albert: and Spinoza's "pantheism",
    103, 115–16; on the Theory of
    Relativity, 10, 113, 144, 151
electricity 136, 144
elements, the four **12**, 26, 29–30, 159
emanation: Augustine's opposition to
    45; Bonaventure's use of 54, 55, 56;
    Greco-Roman idea of, 23, 37; Plotinus
    on 34–6; Erigina on, 47, 48; Silvestris
    on, 68, Bacon's use of 85
End of Nature, The (McKibben) 154, 155
English Civil War 96, 104, 132
Enlightenment 52, 95 (Golinski), 117, 143
Enneads (Plotinus) 34–5, 40n.38
entropy: anticipated in Newton's concern
    over decay, 136–37, 139; second law of
    thermodynamics, 140
environmental philosophy and ethics 116–17
Epicurus, 30, 31, 112, 141; *Peri Phuseos* (On
    Nature), 31
epidemics. *See* disease epidemics
Epistemological question 23. *See also*
    Greco-Roman concepts of nature
equilibrium systems 140
Erasistratus 28
Erigena, 4, **12**, 44, 46–7, 58–9n.12, 73; basis
    for pantheism in (and condemned by

the Church for), 46, 47, 58–59n.12;
and Bruno, 73, 109; on the creating
process, 47–9; on the created world
(defined), 48; on the created world
(formation of), 49; on the First
Principle, 49; on God as creating, but
not created, 46; ; influence of Plotinus
on, 47, 50; *Periphyseon* (*Concerning
Nature/The Division of Nature*),
47–50; on the return to God/Nature,
49–50; on the system of "nature
creating" (*natura creans*) and nature
created (*natura creata*), 47–8
Ethica (*The Ethic*—Spinoza), 104–6,
110–11, 129
Ethical question 23–4. *See also*
Greco-Roman concepts of nature
ethics, 150, 153: ethical accountability, 162.
See also partnership ethic
Euclidian geometry 14, 97, 106, 112,
117, 134
evolution, Darwinian 7, 10, 154, 164
Experiment on a Bird in the Air Pump, An
(Wright) 93
experimentation, **12**, 25, 113, 158; the
constraint of nature in (Bacon), 83,
85–6, 90; as the inquisition of nature
and (Bacon), 89, 90–91; as "vexing
nature" (*see also* vexation), 81, 82–4,
86, 87. *See also* empirical science
extraction and resource exploitation, 65;
mining and metallurgy, 8, 66
Extended Substance (Spinoza) 107, 111–12

famines 42, 64, 66, 67, 93, 103, 142, 155;
Famine of 1315–1317, 65
female nature, 8, 31–2, 67–8; artistic
depictions of, 70–72; *Khaos* as original
creatrix (Hesiod), **12**, 13, 21–3, 26;
Natura Creatrix (Lucretius), **12**, 22,
31–3; personified as *Natura*, 67–8,
70–73
fermentation 135–36, 139, 141
Ficino, Marsilio 73
fire-resistant vegetation 163
First Principle: in Bonaventure, 54, 56; in
Erigena, 49; in Plotinus, 36
five precepts of the partnership ethic 162
flooding, 2, 11, 42, 72, 93, 141, 156, 163;
flood plains, 163

force: both attractive and compulsive
forces, 136; fundamental concept of
(Newton), 134, 136–37; gravitational
force (Newton), 134; impressed force
(Newton), 134. *See also* living force
(Leibniz)
forest depletion 65, 66
forest fire 163
fossil fuels: burning of (carbon emissions),
5, 6, 149–50, 154–55; moving to other
energy sources, 163–64
four causes, the: Aquinas on, 51; Aristotle
on, 29–30
four elements, the (Plato) **12**, 26,
29–30, 159
Fowler, Thomas 85
Fox version of the Bible 88
free will 45, 56

Gaia (goddess) 23
Galileo Galilei 96
Garrard, Mary, on Leonardo da Vinci,
70–72
gases, motions of particles in, 151
Gassendi, Pierre, 30; *Physics*, 105
Gauss, Carl Friedrich, 4, 9, 151
Geb (Egyptian god), 8
gendered nature, 11, 14; in early Greek
thought, 22–3, 25–6, 37; personified as
female (*see also Khaos; Natura*; Natura
Creatrix), 8, 31–2, 67–8. *See also*
female nature
gender equality 162, 163
Genesis, Book of. *See* Bible
Gibbs, J. Willard 9, 151
Glauber, Johann Rudolph 98–9n.14
Gleick, James, *Chaos: Making a New
Science*, 151–52
Global Crisis (Parker) 94
global temperature 155. *See also* climate
change
Glover, Mary, and vexation, 81, 82
God's relation to nature: as intellectual
and non-interventionist (Leibniz),
126, 130, 137–38; manifest in active
principles (Newton), 135–36, 139–40;
monistic pantheism of Spinoza,
108–12, 114, 117, 129; as unity of
creator and creation (Erigena),
46–47, 48

188 *Index*

Goethe, Johann Wolfgang von 115
Goldstein, Herbert S. 116
Golinski, Jan 95
Gomperz, Theodor, on the Hippocratic authors, 28, 28–9, 38n.15, 84
good and evil 129
government 105. *See also* civil society
Grant, Hardy, "Leibniz and the Spell of the Continuous", 130–31
gravity, law of (Newton), **12**, 132, 134; as an active principle of God, 136, 139, 141 cause of, 139; the gravitational constant, 134; and irregularities in nature, 135
great chain of being 159
"*Great Lady, The*" (Leonardo) 71
Great Plague of London (1665) 132, *133. See also* Black Death outbreaks
Greco-Roman concepts of nature, **12**, 13, 21–3; atomism, **12**, 30; on change, 23, 25, 26, 30; early questions, 23–5
Greenland interior 154
"green science" (Soper) 159–60
Guerike, Otto von, air pump, 91–2

Hacking, Ian 89, 166n.5
Hague, The 104, 125, *126*, 128, 129
Halley, Edmund 94
Hayles, Katherine, *How We Became Posthuman*, 160–61
Heaven: empyrean heaven, 12t, 55, 68; *Heaven of the Philosophers* (Paracelsus) 87; in the medieval view of the cosmos, **12**; as Ouranos (male), 23
heavens, 32, 36, 55, 68, 74, 85, 117; *On the Heavens* (Aristotle), 50
Hegel, Georg 47
Heisenberg, Werner, Uncertainty Principle, 10, 144
Heraclitus, 9, **12**, 38n.5; on change and uncertainty in nature, 24–5, 26
Herophilus, 28
Hesiod, **12**; on Khaos, 21–22; *Theogony*, 67
Hippocrates 27–29
Hippocratic treatise: Gomperz on, 28, 28–9, 38n.15, 84; *On the Techne* (On the Art), 27–8, 84
History of Philosophy (Weber) 74, 76n30
history of science, 1, 2, 14, 24; efforts to predict and control nature in, 13,

31–2, 154; prehistory of chaos theory, 6, 13, 23
"History of Winds" (Bacon) 94
Hobbes, Thomas, 30, 84, 96; *Leviathan*, 112
Hooker, Richard, *Of the Laws of Ecclesiastical Polity*, 72
Hooke, Robert, 94; Hooke's Law on earthquakes, 141, 147n.40; *Lectures and Discourses of Earthquakes*, 141
Hoover, Herbert 164
How We Became Posthuman (Hayles) 160–61
human and nonhuman communities 45–6, 162
human body and nature's body 84–5
human habitats 11–12
human-nature connection 154
Human Power and Human Knowledge (Bacon) 85
human soul, Augustine on, 45–6
Hume, David 143
Huxley, T.H. 115
hydrodynamics 144

idealized models of nature, 126–27, 144–45n.4, 146n.4; mathematics and (*see also* mathematics), 134. *See also* pure forms
ignorance of nature 143
Industrial Revolution 6;
industrialization, 140, 144, 155, 156
Inquisition, medieval 88–9
inquisition of nature (Bacon), 89, 90–91; *Folterzwang* ("coercion by torture"), 28
Instauratio Magna (Bacon), 28
Intellect (*Nous*), **12**, 109; Plotinus on, 22, 23, 25–26, 34–5, 36, 41n.39
intellectual and non-interventionist God (Leibniz), 126, 130, 137–38
irreversibility 7
Italian city-states 65
Ivanpah solar array 163–64

Jacobi, Friedrich Heinrich, 114–15
Jonson, Ben, *The Alchemist*, 88
justice, the partnership ethic and, 164–65

Kautz, Richard, *Chaos: The Science of Predictable Random Motion*, 10–11

Kepler, Johannes 96
Khaos, as the original creatrix (Hesiod), **12**, 13, 21–3, 26. *See also* chaos
Krüger, Lorenz, "The Slow Rise of Probabilism", 166n.5
Kubrin, David, xiii, 100n39, 147n40
Kuhn, Thomas, *The Structure of Scientific Revolutions*, 2, 150–51; 165n.5;

Laboratory Life (Latour) 157
labyrinth of the continuum (Leibniz) 130
LaPlace, Pierre Simon, 4, 117, 146n.21; *Analytic Theory of Probabilities*, 9, 151
Latour, Bruno: critique of, 158–59; *Laboratory Life*, 157; on natures-cultures, 157–58, 167n.30; *We Have Never Been Modern*, 157–59
Laws of Ecclesiastical Polity, Of the (Hooker) 72
laws of nature, 22, 97, 125; as the basis for polity (Hooker), 72; cross-cultural aspect, 158–59; Enlightenment interrogation of (*see* Scientific Revolution (Seventeenth c.); Bacon, Francis); knowledge of (in Leonardo da Vinci), 71; *Nomos* (law), 13t., 24; Spinoza on the, 103, 110
least squares, theory of (Gauss), 151
Lectures and Discourses of Earthquakes (Hooke) 141
Leibniz, Gottfried Wilhelm, **12**, 14, 84, *127*; discovery of calculus, 125, 128, 129–30, 131; "Dissertation on the Art of Combinations", 129; on an intellectual and non-interventionist God, 126, 130, 137–38; on the labyrinth of the continuum, 130; on the law of noncontradiction, 131; on the law of sufficient reason, 131; on self-acting monads, 130, 131–32; Spinoza and, 103, 104, 125, 128–29; "The Monadology", 131; *Theodicy*, 131; "The Principles of Nature and Grace", 131; on *vis viva* or mv^2 (idea of kinetic energy), 130–31
"Leibniz and the Spell of the Continuous" (Grant) 130–31
Leibniz-Clarke correspondence 137
Le Monde (*The World*—Descartes) 106

Lessing, Gotthold Ephraim, 114
Leviathan (Hobbes) 112
life after death **12**
lightning 3, 90
Lisbon Earthquake (1755), 11, **12**, 127–28, 140–43, *142;* active intervention by God and, 139
Little Ice Age, 65
living force (Leibniz) 130, 131
Locke, John, 94, 96; *Two Treatises on Government*, 72
London, great fire of 1666, 132
Lorenz, Conrad, **12**; "Does the Flap of a Butterfly's Wings in Brazil Set off a Tornado in Texas?", 152
Lucretius (Carus, Titus), 30, 102, 112, 114, 141; on causes of earthquakes, 141; *De Rerum Natura* (On the Nature of Things), 31; on Natura Creatrix, **12**, 22, 31–3

Magee, Joseph 52
Malagrida, Gabriel 141–42
market economy, in the Middle Ages, 65
Marris, Emma, *Rambunctious Garden*, 153–54
material substratum (*hyle*) 25
mathematical predictability, 130–32, 144; as an antidote to chaos, 125; closed systems and, 126–27
mathematics, **12**, 14, 25, 106, 113; advancing the control and predictability of nature, 131, 144; certainty in, 72; cross-cultural aspect of, 158–59; as a description of nature, 130; Enlightenment models of, 84, 97, 112–13, 117; idealized modeling of nature in, 134; stochastic (*See also* probability theory), 9
matter-force dualism 134
matter in motion: Descartes on, **12**, 106–107, 112; Newton on, 134
Mauna Loa Observatory 6
Maxwell, James Clerk 9, 146n.21, 151
McKibben, Bill, *The End of Nature*, 154, 155
meanings of nature 1, 8–11, **12**
mechanics, Newtonian, 2, 7, 11, 95–96, 132–37; fundamental concept of force, 134, 136–37; impenetrable particles separated by void space, 135

190 Index

mechanism: advances and limitations of, 126–27, 132, 144, 153, 155; as deterministic mechanics (*see also* determinism), 9–10, 140, 146n.21; eliminating qualitative explanations, 134; machine-like model and metaphor of nature in, 137–139; the mechanistic paradigm, **12**, 150, 157; reductionism in (*see also* idealized models of nature), 134, 146n.21; rejecting ideas of chaotic aspects of nature, 130, 131, 135, 153

Meditations on the First Philosophy (Descartes), 132

Megenberg, Konrad von, *Buch der Natur* 57

Mendel, Gregor 10

Merchant Carolyn, *Death of Nature*, xii, 76n4, 100n39; *Ecological Revolutions*, xii; *Earthcare*, xii, 7–8, 15n19; *Radical Ecology*, 168n42; *Reinventing Eden*, xii, 168n42

Merton, Thomas, *natura naturans* in, 149, 150

Middle Ages: natural disasters occurring in, 42–44, 64–5, 66; population levels, 65; preindustrial capitalism in, 66–7; shift from manorial to market economy, 64–6; three-field system, 65. *See also* scholastic philosophy (Medieval)

Mill, John Stuart 162

mind. *See* Intellect (*Nous*); rationality

mining and metallurgy 8, 66

Minoan volcanic disaster 1, **12**, 14n.1

minorities, inclusion of, 162

miracles 86, 107–8, 139, 143, 146–47n.35

Mitchell, Sandra: "Case for Complexity, The", 5–6; *Unsimple Truths*, 5–6

modes or states of Substance (Spinoza) 110, 111–12, 113, 129

Mojave Desert, solar plant in, 163–64

Monad (the One): Being as One (Plotinus), **12**, 25, 34, 36; God as Oneness (Cusa), 69. *See also* Demiurge (Plato)

monads, self-acting (Leibniz) 130, 131–32

monism, pantheistic (in Spinoza), 102–3, 109–12, 113–16. *See also* Monad (the One)

moral consideration 162

motion, matter in, **12**, 112, 131: Descartes' law of motion, **12**, 106–7, 112; ever

new motions needed to offset decay (Newton), 136–37; Newton's laws of, 132, 134

Mt. Vesuvius volcano (Italy 79 C.E.) 11, **12**, 21, *22*

multi-leveled complexity 5

multiple universes, **12**

Naddaf, Gerard 38n.4

Naess, Arne 116, 117

Natura: artistic conceptions of, 70–1; as God's instrument (Hooker), 72–3; as personified goddess (*see also* female nature), 67–8

natura creans/natural create (creating and created world—Erigena), 44, 48

Natura Creatrix, of Lucretius **12**, 22, 31–3

natural disasters: Aquinas on, 52; attributed to Nature's rebellious actions, 67; attributed to corruption of matter (Aquinas), 52; during the Middle Ages, 42–4, 64–5, 66; as God's retribution for sinfulness, 42, 44, 57, 63, 67, 141–42; human (anthropogenic) impacts and, 11, 13, 143, 163; revealing the limits of mechanistic science, 127; signifying the autonomy of nature (*see also* nature as autonomous; unruly and uncontrollable nature), 21. *See also* by type or by disaster

natura libra (Lucretius) 32

natural knowledge 90–93

natural law, 97; applied to civil society, 96; in Aquinas, 52–53; ignorance of or unexplained, 139, 143; prescriptive aspect, 95; Spinoza on (*see also* Spinoza, Baruch), 101, 104–5, 112

natura naturans (nature naturing; nature creating) xii, 8–9, 25, 52, 56; as an active power (Aquinas), 50–51, 53; Bonaventure on, 54, 56; as creative force, 73, 74, 93, 97; Merton on, 149, 150; natural law as a reflection of (Spinoza), 53. *See also* nature as autonomous; unruly and uncontrollable nature

natura naturans/natura naturata: as conceptions of the relationship of creator to creation, 8–9, **12**, 13, 23, 25,

31, 37, 44, 50–51, 56, 70–71; creator and creation as the same (Spinoza), 105, 108–12, 117, 122n.33; creator as distinguished from creation, 14, 31, 62n.39; nature free, in error, or in bonds (Bacon), 83, 85–6, 90, 95–6; tensions between and dynamic interactions, 150, 156–57, 164–65; as unified (contemporary), 161; use of both terms by Bonaventure, 50, 54, 56, 57, 61n.38

natura naturata (nature natured, created world), 8–9, **12**, 13, 23, 50, 53, 54, 61n37, 67, 97, 110–11. *See also* phenomenal world

natura non creata, sed cretatrix, God as the uncreated creator (Augustine), 45–8, 50–51, 54, 56

nature as autonomous, xi, xii, 7–8, 11, 13, 14, 16n.19, 21, 31, 165; chaos theory rooted in perception of, 149, 150, 153, 161; climate change and (*see also* climate change), 149–50, 154–55, 161; early conceptions of (*see* unruly and uncontrollable nature); envisioning a world without humans, 155–57; human partnership with (*see* partnership ethic); new paradigms of nature and, 150, 157, 165; in the partnership ethic, 153, 161–64; therefore unpredictable (*see* uncontrollable nature; unpredictability of nature)

nature as female. *See* female nature

nature naturing/nature natured, 8, 14, 44, 50, 54, 67, 144, 159; Bacon on 84–86; Bonaventure on, 54, 56, 57, 61n37–39; Spinoza on, 102, 109–11, 112, 113, 117

natures-cultures (Latour) 157–59, 161, 167n.30

Necessity: *ananke* (Plato), 26, 38n.10; as arising from natural law (Spinoza), 111

Neoplatonism, 33, 47, 86; Christian Neoplatonism, 44–5, 47, 69, 71, 73. *See also* Christian Neoplatonism; Plotinus

Netherlands, 66

New Atlantis, 90, 94

Newton, Isaac, 30, 84, 96, *128*; on active principles reinvigorating the world, 126; concern that the universe is running down, 136–37, 139, 140; God is present through active principles, 126, 135–37, 139–40; Newtonian mechanics (*see also* mechanism), 2, 7, 11, 95–6, 132–37; on irregularities in nature, 135; theory of optics, 132, 135–36; *Opticks*, 135–36; *Principia Mathematica*, 84, 132. *See also* Scientific Revolution of the Seventeenth c.

Newtoq, Alaska 149

New York City 156

Nomos (law) 13t., 24

noncontradiction, law of (Leibniz) 131

nondeterministic systems. *See* complex systems; unpredictability

non-human nature, 159–60, 161; and human communities, 45–46, 162; inclusion of, 162

non-human (Soper), 159–60

non-interventionist God (Leibniz) 126, 130, 137–38

nonlinearity 7

non-modern world (Latour) 158

Nous. *See* Intellect (*Nous*)

Novum Organum (Bacon) 28, 84, 85

nuclear apocalypse 14

Oldenburg, Henry 104, 129

Oneness. *See* Monad (the One)

On Law and Natural Law (Aquinas) 52

"*On Miracles*" (Spinoza) 107–8

"On Nature, Contemplation, and the One" (*Ennead 3*—Plotinus) 34–5

On the Techne (On the Art—Hippocratic treatise) 27–8, 84

On the Trinity (Augustine) 44

open spaces 163

open systems 140. *See also* complex systems

Opera Postuma (Spinoza) 129

Opticks (Newton) 132, 135–36

order and disorder: disorder in, 14, 140; order arising from disorder (Prigogine), 140; tensions between, 32, 67, 95, 105

Orpheus (Greek god) 8

Other, nature as 159

Pan (Roman god of the forest) 8

pantheism, 115–16; in Bruno, 73–4; Erigena accused of, 46, 47, 58–9n.12;

192   *Index*

pantheistic monism in Spinoza, 102–3, 109–12, 113–16; Toland on, 114–15, 122n.40

Paracelsus, on alchemy, 86–8, 98–9n.13

paradigms in science, 2, 10, 25, 143, 150–53; Botkin on, 5, 6; chaos and complexity (*see also* chaos theory), xi, 1, 2, 3, 4, 6, 8, 13, 14, 144, 151, 153, 156–57, 160–61; climate change and, 1, 6, 150, 151; Deep Ecology and, 116; Kuhn on, 2, 150–51, 165n.5; Mitchell on, 5, 6; the mechanistic paradigm, **12**, 150, 154, 157; ew paradigm for 21st century (Merchant), xi, 2, 6, 7, 150–53, 158; shift toward a sun-centered cosmos, 96–7, 151; succession of paradigms in ecology (Simberloff), 9–10, 16n.26; Wells on, 3, 6. *See also* nature as autonomous; probability theory; scientific revolutions

Paris Academy of Sciences 91

Parker, Geoffrey, *Global Crisis* 94

Parliament of Things (Latour) 158

Parmenides, 9, **12**, 38n.5; on the permanence of nature, 24

particles: atomism and, 113, 132; attractive and repulsive forces acting on, 136; as hard masses separated by void space (Newton), 135; nature consisting of, **12**; random motions of, 151

partnership ethic, xii, 14, 15n18; climate change and, 153, 164–65; combining utilitarian and ecocentric ethics, 161–63; five precepts of, 162; nature as autonomous in, 153, 161–64; practical implementations of, 163–64

Peri Phuseos (On Nature—Epicurus) 31

Periphyseon (*Concerning Nature/The Division of Nature*—Erigena) 47–50

permafrost, melting of, 19

permanence vs. change 25

personified nature, during the Renaissance 64

perspective art 71–2

Petrarch, Francesco 42

phenomena bene fundata (Leibniz) 131

phenomenal world, 31–2, 73, 74, 75, 109–10, 126, 130–31; as empirical world, 29; as modes or states of

Substance, 110, 111–12, 113, 129; as self-acting monads (Leibniz), 130, 131–32; as the world of appearances, 31–2, 73, 74, 75, 109, 126, 130–31. *See also natura naturata (created world)*

physis (*phusis*), 8, **12**, 24; as the material principle (Silvestris), 68; and *techne* (Hippocrates), 27–9

physis/natura, in the Hippocratic Treatise, 84

Physis (*phusis*), (*Physics*—Aristotle) 24, 25, 29, 39, 43, 50, 51, 67

Pilkington, Ed 149

Planck, Max, on quantum theory, 10

Plato, **12**, 23, 29, 33; on the Demiurge, 22, 23, 25–27, 68, 134; on Necessity (*ananke*), 26, 38n.10; Plotinus following, 33, 34; on pure forms, 25, 26, 29, 34, 55, 134; *Timaeus*, 25, 67

Platonic Form 9

Pliny the Elder 21

Pliny the Younger 21

Plotinus, **12**, 22, 29, 86; the *Enneads*, 34–35, 40n.38; following Plato, 33, 34; influencing Medieval theologians, 45, 46, 47; on Intellect (*Nous*), 22, 23, 25–6, 34–5, 36, 41n.39; on the Monad (First Principle), **12**, 25, 34, 36; on the Soul (*Anima*), **12**, 34–5; on *Tolma* (rebellious audacity of the World Soul), 34–7, 40n.38, 46, 68, 73, 86, 95, 154. *See also* Neoplatonism

Plumptre, Constance E. 123n45

politics, new forms of 150

post-human world 155–57, 160–61

post-wild world 153–54

predictability/prediction: deterministic science and (*see also* Newtonian mechanics), 2–5, 76, 84, 90, 95, 97, 127; of the laws of nature, 95; the problem of (*see also* chaos theory), 152–53; the search for (*see also* mathematical predictability), 125. *See also* certainty, problem of

preindustrial capitalism 66–67

Prigogine, Ilya, **12**, 140; and Isabelle Stengers *Order out of Chaos*, 7, 10; on the problem of predictability, 152–53

Primoridal (*Silva*) 68

Principia Mathematica (Newton) 84, 132
Principiis Naturae ("On the Principles of Nature"—Aquinas) 51
Principles of Descartes's Philosophy (Spinoza) 105, 106–107, 110
Principles of Philosophy (Descartes) 105, 135
probability theory, 4, 9, 151, 166n.5; the law of large numbers, (Bernouli; LaPlace), 151; statistics, 4, 9, 151; stochastic mathematics and, 9
process-orientated perspective 9–10
Protestant Reformation 96
Proteus, myth of (Bacon) 28, 82, 87, 88
psychology 150
Ptolemy, earth-centered cosmos in, 96, 151
pure forms: Bonaventure on, 55–6; in Erigena, 48; mathematics and, 134; Plato on, 25, 26, 29, 34, 55, 134; vs. the world of appearances, 9, 23, 25–6, 34, 160. *See also* idealized models of nature

qualitative explanations 134
quantification, 12; quantities as primary qualities, 134. *See also* mathematics
quantum theory, 10, 144, 151–52; quanta/quarks, 12

Rambunctious Garden (Marris) 153–54
randomness, 4–5, 9, 10–11, 144–45n4, 160–61; random motion (Lucretius), 31, 32, 33; statistics and, 151. *See also* chaos; uncertainty
rationalism, 126–27, 131; as the foundation of natural philosophy, 126
rationality: Christianity philosophy of, 44, 45, 49, 51, 52, 56, 57; of nature, 14, 30, 32, 33, 38n.4, 86, 95, 97; voluntarism and, 73; in the work of Aquinas, 44, 45, 49, 51–3, 56
rebellious nature, 33–7, 67, 95; audacity of the World Soul (Plotinus), 34–7, 40n.38, 46, 68, 73, 86, 95, 154; audacity of the World Soul (*Tolma*), 40n.38, 46, 68, 73, 86, 154; in Leonardo, 72
Rees, Graham 87
relativity theory (Einstein), 10, 113, 144; technological advances based on, 151

Renaissance: artists of, 70–71; meanings of nature in, 12, 13
Resurrection, Christian, 49–50
Righteous Persecution (Ames) 88
Rohr, Christian 57
Rome, ancient, 12, 13
Rousseau, Jean-Jacques 143
Royal Society of London 91, 104

Salomon's House experiments 90
salvation 12
Sannig, Bernard 56
Santorini volcano 14n.1
Satan: and fallen angels, 46, 58n.9; possession by, 81, *82. See also* free will; sinfulness
scholastic philosophy (Medieval), 12, 37; Christian Neoplatonism in, 44–5, 47, 69, 71, 73. *See also* Aquinas, Thomas; Augustine; Erigena
Scientific Revolution of the Seventeenth c., 2, 4–5, 95, 165; deterministic mechanics and, 9
meanings of nature and, 12; rise of mathematical models (*see* mathematics); transition to, 73–5. *See also* Bacon, Francis; Newton, Isaac; and other scientists by name
scientific revolutions: Cohen on, 165–66n.5; Hacking on, 89, 166n.5; Kuhn on, 2, 150, 165n.5; models of probability the 18th and 19th c., 151; Twentieth c. science, 2, 10, 12 *See also* chaos and complexity; quantum theory; relativity theory
Scotus, Johannes (John) 37, 46, 58n.10
sea level rise 155, 156. *See also* climate change
sense perception 29. *See also* world of appearances
sensuality 53
Sessions, George 116
Short Treatise on God, Man, and His Well-Being (Spinoza) 104, 109, 110
Silvestris, Bernard, *Cosmographia* 68–9
Simberloff, Daniel, "Succession of Paradigms in Ecology" 9–10, 16n.26
sinfulness, 46, 56; and disaster, 42, 44, 57, 63, 67

194 *Index*

"Slow Rise of Probabilism, The" (Krüger) 166n.5
social class 25, 66, 177
social construction of natural disasters 11, 17n.30
social systems 140
societies-natures. See natures-cultures
sociology of knowledge 157
solar-based world 163–64
Soper, Kate, *What Is Nature?* 159–60
soul: human, 45–46; Soul (*Anima*) of Plotinus, **12**, 34–5; Soul of the world—medieval (*see* World Soul)
Special Relativity, Theory of (Einstein) 10
species annihilation 155–57. *See also* climate change
Spedding, James B., Robert Ellis, and Douglas Heath, 85
Spinoza, Baruch, **12**, 13, 84, 86, 97, *102;* on Cartesian dualism, 103, 105–7, 120n.15; on civil society, 104–105; *Ethica (The Ethic)*, 104, 105, 106, 110–11, 129; on God and Nature as One (*Deus sive Natura*), 108–12, 114, 117, 129; on the laws of nature, 104; on modes or states of Substance, 110, 111–12, 113, 129; on natural law, 101, 104–105, 112; on natural law and *natura naturans*, 53; "On Miracles", 107–08; *Opera Postuma*, 129; personal life, 101, *102*, 103–04, *126; Principles of Descartes's Philosophy*, 105, 106–07, 110; rational system of philosophy, 101–03, 118n.1; *Short Treatise on God, Man, and His Well-Being*, 104, 109, 110; on Substance and Extended Substance, 107, 111–12, 113; that the creator (*natura*) and creation (*naturata* are the same, 105, 108–12, 117, 122n.33; on Thomism, 110; *Tractatus Theologico-Politicus*, 104, 105, 106, 128
Sprat, Thomas 92–93
state of nature: Hobbes on, 96; Spinoza on, 104
statistics 4, 9, 151. *See also* probability theory
steam engine 154–55
Stellar Sphere (*De Caelo*) **12**

Stewart, Matthew 125
stochastic mathematics, 9. *See also* probability theory
storms 11, 42, 66, 72, 135, 156
Structure of Scientific Revolutions, The (Kuhn) 2, 150–51
"Succession of Paradigms in Ecology" (Simberloff) 9–10, 16n.26
sufficient reason, law of (Leibniz) 131
Summa Theologica (Aquinas) 51–2, 109
sun-centered cosmos, paradigm shift toward, 96–7, 151. *See also* Copernicus, Nicholas
sustainability 3
Swan, John 97n.1–2
techne: human arts and the constraint of nature (Bacon), 83, 84, 86; as human arts and the constraint of nature (Hippocrates), 27–29; as human art/technology (Cusa), 69–70. *See also* Hippocratic treatise

technology, 69–70, 144; advances of relativity theory in, 151. *See also techne*
telos (goal—Aristotle) **12**, 30
Thema Coeli (Bacon) 87
Theodicy (Leibniz) 131
Theogony (Hesiod) 67
theology, freeing nature from, 91, 99–100n.28
theory of optics (Newton) 132, 135–36
Theory of Relativity (Einstein) 113, 144, 151
Theory of Special Relativity (Einstein) 10
thermodynamics, 140, 144; far-from-equilibrium systems, 140; second law of (entropy), 140
1348 Earthquake (Italy), **12**, *43*
Thomists 110. *See also* Aquinas, Thomas
"350.org" 155
three-field system 65
three states of nature (Bacon): first state, 95; second state, 86, 87, 95; third state, 86, 87, 95–6
Tiles, Mary 89–90
Tillyard, Eustace 73
Timaeus (Plato) 25, 67
Timescapes of Modernity (Adam) 133–34

Toffler, Alvin 10
Toland, John, 114–15, 122n.40
Tolma, as rebellious audacity of the World
  Soul (Plotinus), 34–7, 40n.38, 46, 68,
  73, 86, 95, 154
Torricelli, Evangelista 132
Tractatus Theologico-Politicus (Spinoza)
  104, 105, 106, 128
transmutation, alchemical 87–8. *See also*
  alchemy
Trinity of God, Bonaventure on, 54–6
Tschirnhaus, Walther Ehrenfried von
  104, 128
tsunamis 1, 11, 14n.1, 141
21st Century scientific perspectives **12** *See
  also* chaos and complexity; scientific
  revolutions
Two Treatises on Government (Locke) 72

uncertainty, 1, 4–5, 6, 26, 161; change
  and uncertainty in Heraclitus, 26; in
  chaotic systems, 10
  principle of (Heidegger), 10, 144; *See
  also* complex systems; randomness
unpredictability: due to God's presence
  in active principles (Newton), 133,
  135–36, 139–40; levels of, 4–6;
  rejected by mechanists, 130, 131, 135;
  and uncontrollability (*see also* unruly
  and uncontrollable nature), 1, 2–4,
  14, 22, 36, 150. *See also* chaos theory;
  natural disasters
unruly and uncontrollable nature: chaos
  theory and, 149, 150; disorder in,
  14, 140; *natura libra* (Lucretius),
  32; *natura naturans* as creative
  force, 73, 74, 93, 97; nature as
  uncontrollable force, 8, 155, 156;
  uncontrollable events (*see* climate
  change; natural disasters; weather
  forecasting); the unpredictability of
  (*see* unpredictability), 1, 2–4, 14, 22,
  36, 150; as unruly nature, 13, 65, 66,
  67, 93–5, 101, 125. *See also* chaos;
  control of nature; natural disasters;
  unpredictability
Unsimple Truths (Mitchell) 5–6
unstable systems 152–53. *See also* complex
  systems

Urania, as the celestial principle
  (Silvestris), 68
utilitarian ethic 161–62, 165

Venus 32
Vesuvius, Mt. volcano (Italy 79 C.E.) 11,
  12, 21, *22*
vexation: alchemical references to, 88, 89;
  Biblical use of term, 88; in the myth
  of Proteus (Bacon), 28, 82, 87, 88;
  of nature through experimentation
  (Bacon), 81, 82–4, 86, 87; as Satanic
  possession in witchcraft, 81, 82, 86;
  torturing as a form of, 28, 29, 39n.17,
  88–9
Virgin of the Rocks (Leonardo da Vinci) 71
vitalism, *vis viva* in Leibniz, 130–31
volcanoes, 42, 66; Crete volcanic disaster
  1, **12**, 14n.1; Mt. Vesuvius volcano
  (Italy 79 C.E.), 11, **12**, 21, *22;* Santorin
  volcano, 14n.1
Voltaire, François-Marie A., 142–43
voluntarism: of God in Augustine, 45; and
  a non-interventionist God (Leibniz),
  126, 130, 137–38; rationality and, 73
von Staden, Heinrich, 27, 28, 84

Watt, James, 154
weather forecasting, 6–7, 93–5
Weber, Alfred *History of Philosophy*, 74,
  76n30
We Have Never Been Modern (Latour) 157–59
Weiner, Norbert 160
Weisman, Alan, *The World Without Us* 155–56
Wells, Jennifer, *Complexity and
  Sustainability* 3, 152–53
Werndl, Charlotte 2
Western cultures and non-Western cultures
  157–59
Williams Raymond, on the meaning of
  nature, 6, 8
wisdom of God (Leibniz) 138–39
Wisdom of the Ancients (Bacon) 28–9
witchcraft, 97n.2; the Inquisition and,
  88–89; Satanic possession, 81, 82, 86
women, inclusion of in ethical
  accountability, 162
world of appearances, 9, 23, 25, 26, 34, 160;
  as the phenomenal world, 31–2, 73,

74, 75, 109, 126, 130–31; versus pure
forms, 9, 23, 25–6, 34, 160
World Soul: Medieval, **12**; Nature as lower
form of, 67; of Plotinus (*Anima*),
12, 34–5, 46; as *psyche* (Plato), 2, 25;
Silvestris on, 68

World Without Us, The (Weisman) 155–56
Worsham, Lynn and Gary Olson, 161;
Plugged In, 161
Wright, Joseph, "An Experiment on a Bird
in the Air Pump" 93
Zwangsmittel (coercive means) 28